GEOMETRY OF SEMILINEAR EMBEDDINGS

Relations to Graphs and Codes

GEOMETRY OF SEMILINEAR EMBEDDINGS

EMBEDDINGS

Relations to Graphs and Codes

Mark Pankov

University of Warmia and Mazury, Poland

World Scientific

NEW JERSEY · LONDON · SINGAPORE · BEIJING · SHANGHAI · HONG KONG · TAIPEI · CHENNAI

Published by

World Scientific Publishing Co. Pte. Ltd.

5 Toh Tuck Link, Singapore 596224

USA office: 27 Warren Street, Suite 401-402, Hackensack, NJ 07601

UK office: 57 Shelton Street, Covent Garden, London WC2H 9HE

Library of Congress Cataloging-in-Publication Data
Pankov, Mark.
   Geometry of semilinear embeddings : relations to graphs and codes / by Mark Pankov
(University of Warmia and Mazury, Poland).
      pages cm
   Includes bibliographical references and index.
   ISBN 978-9814651073 (hardcover : alk. paper)
   1. Embeddings (Mathematics) 2. Geometry, Algebraic. 3. Grassmann manifolds. I. Title.
   QA564.P36 2015
   516.3'5--dc23

                                                              2015008649

British Library Cataloguing-in-Publication Data
A catalogue record for this book is available from the British Library.

Printed in Singapore

To the memory of my teacher
Vladimir Sharko
(25.09.1949 – 7.10.2014)

Preface

The main objective of this book is to expound on semilinear embeddings of vector spaces over division rings and the associated mappings of Grassmannians. A large portion of the material will be formulated in terms of graph theory. Some relations to linear codes will be described. This book consists of results obtained by different mathematicians and published in original research articles only. The author puts together these contributions on the basis of a unified idea.

A *semilinear* mapping $l : V \to V'$ between left vector spaces V and V' over division rings R and R' (respectively) satisfies the following conditions:

$$l(x + y) = l(x) + l(y)$$

and

$$l(ax) = \sigma(a)l(x)$$

for a certain mapping $\sigma : R \to R'$. In the classical notion of a semilinear mapping, σ is an isomorphism of the division ring R to the division ring R' [Artin (1957); Baer (1952); Dieudonné (1971)]. Following [Faure and Frölicher (2000)] we will suppose that σ is a homomorphism of R to R'. Note that non-zero homomorphisms of division rings are injective.

Semilinear isomorphisms (semilinear bijections over isomorphisms of division rings) play an important role in the classical Projective Geometry and Linear Group Theory. For example, the Fundamental Theorem of Projective Geometry states that all collineations (isomorphisms) between projective spaces are induced by semilinear isomorphisms of the corresponding vector spaces [Artin (1957); Baer (1952); Dieudonné (1971)] and there is a modern version of this result concerning a larger class of semilinear mappings [Faure and Frölicher (1994); Havlicek (1994); Faure and Frölicher (2000)]. Also, semilinear isomorphisms are exploited in the description of automorphisms of classical groups [Dieudonné (1971); O'Meara (1974)]. Some structural properties of semilinear isomorphisms can be found in [Dempwolff (1990, 1999); Dempwolff, Fisher and Herman (2000); Dempwolff (2010)].

We will investigate *semilinear embeddings* over not necessarily surjective homomorphisms of division rings. *Strong semilinear embeddings* transferring any collection of linearly independent vectors to linearly independent vectors are similar to semilinear isomorphism. There exist also *non-strong semilinear embeddings* where the structures are more complicated [Brezuleanu and Rădulescu (1984); Faure and Frölicher (2000); Kreuzer 1 (1998); Kreuzer 2 (1998)].

Other important objects considered in the book are *Grassmann graphs*. The Grassmann graph $\Gamma_k(V)$ associated to an n-dimensional vector space V is the graph whose vertex set is the Grassmannian $\mathcal{G}_k(V)$ consisting of k-dimensional subspaces of V. Two k-dimensional subspaces are adjacent vertices of the graph if their intersection is $(k-1)$-dimensional. Any two distinct vertices of this graph are adjacent if $k = 1, n-1$ and we will suppose that $1 < k < n-1$. The Grassmann graphs corresponding to vector spaces over finite fields are classical examples of distance-regular graphs [Brouwer, Cohen and Neumaier (1989)]. We consider the Grassmann graphs of vector spaces over division rings. The vertex set of such a graph can be infinite.

The first result on Grassmann graphs was the well-known Chow's theorem [Chow (1949)]. It describes the automorphisms of Grassmann graphs similarly to the Fundamental Theorem of Projective Geometry. If $n \neq 2k$ then every automorphism of $\Gamma_k(V)$ is induced by a semilinear automorphism of V. In the case when $n = 2k$, there are also automorphisms of $\Gamma_k(V)$ obtained from semilinear isomorphisms of V to the dual vector space V^*. This statement is closely related to Hua's theorem [Hua (1951)] known as the Fundamental Theorem of Geometry of rectangular matrices [Wan (1996); Šemrl (2014)]. If V is a vector space over a field then the Grassmannian $\mathcal{G}_k(V)$ can be considered as a subset in the projective space associated to the exterior power $\wedge^k(V)$ and there is a nice interpretation of Chow's theorem in terms of semilinear automorphisms of exterior powers [Westwick (1964)].

In [Buekenhout and Cohen (2013); Pankov (2010); Pasini (1994)] Grassmann graphs are presented as one type of the graphs related to buildings [Tits (1974)]. Note that there are analogues of Chow's theorem for polar Grassmann graphs and half-spin Grassmann graphs [Chow (1949); Dieudonné (1971); Pankov (2010)]. These graphs are related to buildings of types $\mathsf{B}_n = \mathsf{C}_n$ and D_n. In this book, we distinguish some properties inherent in Grassmann graphs only.

We consider the Grassmann graphs $\Gamma_k(V)$ and $\Gamma_{k'}(V')$ corresponding to vector spaces not necessarily of the same dimension and establish that every *isometric embedding* of $\Gamma_k(V)$ in $\Gamma_{k'}(V')$ is induced by a semilinear embedding of special type. It must be pointed out that this semilinear embedding need not be strong. Chow's theorem easily follows from our description of isometric embeddings. Also, there are non-strong semilinear embeddings which induce non-isometric embeddings of Grassmann graphs.

The *Johnson graph* $J(n, k)$ is formed by all k-element subsets of an n-element set. Two k-element subsets are adjacent vertices of the graph if their intersection consists of $k - 1$ elements. This graph is a thin version of the Grassmann graph $\Gamma_k(V)$, where $\dim V = n$. For every base of the vector space V the associated *apartment* of $\mathcal{G}_k(V)$, i.e. the set consisting of all k-dimensional subspaces spanned by subsets of this base, is the image of an isometric embedding of $J(n, k)$ in $\Gamma_k(V)$. We describe the images of all admissible isometric embeddings of the Johnson graph $J(i, j)$ in the Grassmann graph $\Gamma_k(V)$ (the Johnson and Grassmann graphs need not have the same indices). Such images are said to be $J(i, j)$-*subsets* of $\mathcal{G}_k(V)$. The description of $J(i, j)$-subsets is similar to the description of isometric embeddings of Grassmann graphs. We use the so-called *m-independent* subsets of projective spaces instead of semilinear embeddings.

Our interest to $J(i, j)$-subsets is motivated by the problem of characterizing apartments in building Grassmannians [Cooperstein (2013)]. We observe the following phenomenon: every $J(n, k)$-subset of $\mathcal{G}_k(V)$ is an apartment only in the case when $n = 2k$ [1]. The significance of $J(i, j)$-subsets follows also from the fact that isometric embeddings of $\Gamma_k(V)$ in $\Gamma_{k'}(V')$ can be characterized as mappings which transfer apartments of $\mathcal{G}_k(V)$ to $J(n, k)$-subsets of $\mathcal{G}_{k'}(V')$. This statement essentially generalizes the description of mappings between Grassmannians transferring apartments to apartments [Pankov (2010), Theorem 3.10].

The notion of a *projective system* [Tsfasman, Vlăduţ and Nogin (2007)] enables us to translate some problems of Coding Theory in terms of Projective Geometry. Roughly speaking, projective systems are collections of not

---

[1]It was first pointed in [Cooperstein and Shult (1997)] that there are isometric embeddings of $J(n, k)$ in $\Gamma_k(V)$ whose images are not apartments. A characterization of apartments in other terms can be found in [Cooperstein, Kasikova and Shult (2005)]. We conjecture that for apartments of polar Grassmannians there is a metric characterization similar to the characterization of apartments in $\mathcal{G}_k(V)$ for $n = 2k$.

necessarily distinct points in the projective spaces over finite fields. Every *linear code* is related to a certain projective system and, conversely, every projective system defines an equivalence class of linear codes. There is a simple geometrical proof of the classical MacWilliams theorem [MacWilliams (1961)] based on this approach [Ghorpade and Kaipa (2013)]. Also, it was observed in [Ghorpade and Kaipa (2013)] that Chow's theorem (more precisely, its interpretation in terms of semilinear automorphisms of exterior powers) provides a description of the automorphism groups of Grassmann codes [Nogin (1996)]. Using projective systems, we can connect $J(i,j)$-subsets of Grassmannians associated to vector spaces over finite fields with some linear codes.

For the reader's convenience, we include some basic of algebra (division rings and their homomorphisms, vector spaces over division rings, exterior powers) and all relevant facts of Projective Geometry and Coding Theory. This makes the book self-contained and accessible for graduate students. Prospective audience includes researchers working in linear algebra, combinatorics and coding theory.

Mark Pankov

Contents

Chapter 1

Semilinear mappings

We investigate semilinear mappings of vector spaces over division rings. In contrast to the classical books [Artin (1957); Baer (1952); Dieudonné (1971)] we do not require the associated homomorphisms of division rings to be isomorphisms. For this reason, the first section is devoted to division rings and their homomorphisms.

We will distinguish the following three types of semilinear mappings:

- semilinear isomorphisms;
- strong semilinear embeddings which transfer any collection of linearly independent vectors to linearly independent vectors (such mappings are similar to semilinear isomorphisms);
- semilinear m-embeddings which send any m linearly independent vectors to linearly independent vectors, where m is an integer between 2 and the dimension of the corresponding vector space.

We show that non-strong semilinear embeddings exist. For any integers $n \geq 3$ and $m > 0$ there is a semilinear n-embedding of an $(n+m)$-dimensional vector space in an n-dimensional vector space. Note that the associated homomorphism of division rings is not surjective.

We consider the mappings of Grassmannians induced by semilinear embeddings. Some of them are isometric embeddings of Grassmann graphs. In this way, we get also non-isometric embeddings. Embeddings of Grassmann graphs will be investigated in Chapter 3. Projective spaces are Grassmannians formed by subspaces whose dimension or codimension is equal to 1. The mappings of projective spaces induced by semilinear embeddings will be exploited in the next chapter in context of a generalized version of the Fundamental Theorem of Projective Geometry.

Every semilinear m-embedding defines a mapping between Grassmannians formed by k-dimensional subspaces if $k \leq m$. For semilinear isomorphisms such mappings are bijective. We ask: is this property characterizing for semilinear isomorphisms? In other words, is an embedding a semilinear isomorphism if one of the associated mappings between Grassmannians is bijective? There is a counterexample, where one of the vector spaces is not necessarily finite-dimensional [Kreuzer 2 (1998)]. The problem is open for the case when both vector spaces are finite-dimensional.

By [Dieudonné (1971)], all automorphisms of the general linear group can be obtained from semilinear automorphisms of the associated vector space and semilinear isomorphisms of this vector space to the dual vector space. Every strong semilinear embedding induces a monomorphism between the general linear groups of the corresponding vector spaces. All homomorphisms of such groups are described only in some special cases [Dicks and Hartley (1991); Zha (1996)]. At the end of the chapter we characterize strong semilinear embeddings in terms of the general linear groups (Theorem 1.2). This characterization is connected to the above mentioned homomorphism problem. Also, it is related to the notion of rigid embedding discussed in Chapter 3.

1.1   Division rings and their homomorphisms

Let R be a non-empty set with additive and multiplicative operations $+$ and $\cdot$, respectively. Suppose that $(R, +)$ is an abelian group whose identity element is 0 and $(R \setminus \{0\}, \cdot)$ is a group whose identity element is denoted by 1. Then $(R, +, \cdot)$ is a *division ring* if

$$a(b + c) = ab + ac \quad \text{and} \quad (b + c)a = ba + ca$$

for all $a, b, c \in R$. The latter conditions guarantee that

$$0a = a0 = 0 \quad \text{and} \quad (-1)a = a(-1) = -a$$

for every $a \in R$. A division ring is a *field* if the multiplicative operation is commutative.

For every division ring R there is the *opposite* division ring R^*. The division rings R and R^* have the same set of elements and the same additive operation. The multiplicative operation $*$ on R^* is defined by the formula

$$a * b := ba.$$

It is clear that $R^{**} = R$. The opposite division ring R^* coincides with R only in the case when R is a field.

Let R and R' be division rings. A non-zero mapping $\sigma : R \to R'$ is called a *homomorphism* of the division ring R to the division ring R' if

$$\sigma(a+b) = \sigma(a) + \sigma(b) \quad \text{and} \quad \sigma(ab) = \sigma(a)\sigma(b) \qquad (1.1)$$

for all $a, b \in R$. In the case when $R = R'$, we say that σ is an *endomorphism* of the division ring R. Since σ is non-zero, the equalities (1.1) imply that

$$\sigma(0) = 0 \quad \text{and} \quad \sigma(1) = 1.$$

Every homomorphism of R to R' is a homomorphism of R^* to R'^* and conversely.

Proposition 1.1. *Homomorphisms between division rings are injective.*

Proof. If a homomorphism $\sigma : R \to R'$ is non-injective then $\sigma(a_1) = \sigma(a_2)$ for some distinct $a_1, a_2 \in R$ and $\sigma(a_1 - a_2) = 0$. Hence $\sigma(a) = 0$ for a certain $a \in R \setminus \{0\}$ and for every $b \in R$ we have

$$\sigma(b) = \sigma(aa^{-1}b) = \sigma(a)\sigma(a^{-1}b) = 0$$

which is impossible (by our definition, homomorphisms are non-zero mappings). □

Bijective homomorphisms are said to be *isomorphisms*. For every division ring R the group formed by all automorphisms of R (isomorphisms of R to itself) is denoted by $\mathrm{Aut}(R)$. Automorphisms of type

$$a \to bab^{-1}, \quad b \neq 0$$

are called *inner*. A division ring has non-trivial inner automorphisms only in the case when it is non-commutative. If R is a non-commutative division ring then isomorphisms of R to R^* (if they exist) are said to be *anti-automorphisms* of R.

Remark 1.1 ([Hua (1949)]). Suppose that $\sigma : R \to R'$ is a mapping satisfying the following conditions:

(1) $\sigma(a+b) = \sigma(a) + \sigma(b)$ for all $a, b \in R$,
(2) $\sigma(a^{-1}) = \sigma(a)^{-1}$ for every $a \in R \setminus \{0\}$,
(3) $\sigma(1) = 1$.

Then σ is a homomorphism of R to R' or a homomorphism of R to R'^*. The following examples show that each of the above conditions cannot be omitted:

- non-additive homomorphisms between the multiplicative groups of division rings (for example, $R = R'$ is a field and $\sigma(x) = x^{-1}$ for all $x \in R \setminus \{0\}$),
- linear transformations of the real vector spaces corresponding to the field of complex numbers $\mathbb{C}$ and the quaternion division ring $\mathbb{H}$ which leave fixed 1 and are not multiplicative preserving,
- $\sigma = -\delta$, where δ is a homomorphism of division rings.

Example 1.1. Every endomorphism of $\mathbb{Q}$ or $\mathbb{R}$ is identity. If $\sigma : \mathbb{Q} \to \mathbb{Q}$ is an endomorphism then

$$\sigma(n) = \underbrace{1 + \cdots + 1}_{n} = n \ \text{ and } \ \sigma\left(\frac{n}{m}\right) = \frac{\sigma(n)}{\sigma(m)} = \frac{n}{m}$$

for all $n, m \in \mathbb{Z} \setminus \{0\}$. Similarly, if $\sigma : \mathbb{R} \to \mathbb{R}$ is an endomorphism then its restriction to $\mathbb{Q}$ is identity. Using the fact that every positive $a \in \mathbb{R}$ is a square, i.e. $a = b^2$ for a certain $b \in \mathbb{R}$, we establish that σ transfers positive numbers to positive numbers which means that σ is order preserving. Then σ sends the Dedekind cut corresponding to $a \in \mathbb{R}$ to the Dedekind cut corresponding to $\sigma(a)$. This implies that σ is identity, since its restriction to $\mathbb{Q}$ is identity.

Example 1.2. The automorphism group of $\mathbb{Q}(\sqrt{2})$ is isomorphic to $\mathbb{Z}_2$. The unique non-identity automorphism sends $a + b\sqrt{2}$ to $a - b\sqrt{2}$.

Example 1.3. Consider $\mathbb{Q}(t)$, where $t \in \mathbb{R} \setminus \mathbb{Q}$ is transcendental. Let σ be an endomorphism of $\mathbb{Q}(t)$. Since its restriction to $\mathbb{Q}$ is identity, σ is completely defined by the value in t which can be an arbitrary element of $\mathbb{Q}(t) \setminus \mathbb{Q}$. An easy verification shows that σ is an automorphism if and only if

$$\sigma(t) = \frac{at + b}{ct + d},$$

where $a, b, c, d \in \mathbb{Q}$ and $ad \neq bc$. In particular, there are non-surjective endomorphisms of $\mathbb{Q}(t)$.

Proposition 1.2. *If F is a subfield of an algebraically closed field E then every automorphism of F can be extended to an automorphism of E.*

Proof. Let σ be an automorphism of F. Consider the family $\mathfrak{F}$ formed by all pairs (K, δ), where K is a subfield of E containing F and δ is an automorphism of K whose restriction to F coincides with σ. For such two pairs (K, δ) and (K', δ') we write $(K, \delta) \leq (K', \delta')$ if $K \subset K'$ and the restriction of δ' to K coincides with δ. Then $(\mathfrak{F}, \leq)$ is a non-empty partially ordered set satisfying the conditions of Zorn's lemma and there is a maximal element (K_m, δ_m). Since any automorphism of a field can be extended to an automorphism of the algebraic closure of this field [Lang (2002), Chapter V, Theorem 2.8] and E is algebraically closed, K_m is algebraically closed. If K_m is a proper subfield of E then every $a \in E \setminus K_m$ is transcendental over K_m. We set $\delta_m(a) = a$ and extend δ_m to an automorphism of the field $K_m(a)$. This contradicts the maximality of (K_m, δ_m). Hence K_m coincides with E. $\qquad\square$

The above statement fails for the case when E is not algebraically closed. Non-identity automorphisms of the fields considered in Examples 1.2 and 1.3 cannot be extended to an automorphism of $\mathbb{R}$.

Example 1.4. The group $\mathrm{Aut}(\mathbb{C})$ is non-trivial. It contains, for example, the conjugate mapping $z \to \bar{z}$. There exist other non-identity automorphisms of $\mathbb{C}$. By Proposition 1.2, non-identity automorphisms of the fields considered in Examples 1.2 and 1.3 can be extended to automorphisms of $\mathbb{C}$. These extensions do not coincide with the conjugate mapping, since the latter is the unique non-identity automorphism of $\mathbb{C}$ leaving fixed all real numbers. By [Yale (1966)], there are non-surjective endomorphisms of $\mathbb{C}$. Let $X = \{x_i\}_{i \in \mathbb{N}}$ be a sequence of complex numbers algebraically independent over $\mathbb{Q}$ and let σ be the endomorphism of $\mathbb{Q}(X)$ transferring every x_i to x_{i+1}. Consider the family of all pairs (K, δ), where K is a subfield of $\mathbb{C}$ containing $\mathbb{Q}(X)$ and δ is an endomorphism of K whose restriction to $\mathbb{Q}(X)$ coincides with σ and such that x_1 is transcendental over $\delta(K)$. As in the proof of Proposition 1.2, we extend σ to a non-surjective endomorphism of $\mathbb{C}$. See [Yale (1966)] for the details.

Example 1.5. Following [Bachman (1964)] we construct the field of p-adic numbers $\mathbb{Q}_p$. Let us fix a real number $c \in (0, 1)$ and a prime integer p. Every non-zero rational number x can be presented in the form

$$x = p^{\alpha} \frac{a}{b},$$

where $\alpha, a, b \in \mathbb{Z}$ and $p \nmid a$, $p \nmid b$. We define

$$|x|_p := \begin{cases} c^{\alpha} & \text{if } x \in \mathbb{Q} \setminus \{0\}, \\ 0 & \text{if } x = 0. \end{cases}$$

The set of rational numbers together with the distance function

$$d(x, y) := |x - y|_p, \quad x, y \in \mathbb{Q}$$

is a metric space. Let $\mathcal{C}$ be the set of all Cauchy sequences in this metric space. For any $\{a_i\}_{i\in\mathbb{N}}, \{b_i\}_{i\in\mathbb{N}} \in \mathcal{C}$ we define

$$\{a_i\}_{i\in\mathbb{N}} + \{b_i\}_{i\in\mathbb{N}} = \{a_i + b_i\}_{i\in\mathbb{N}} \text{ and } \{a_i\}_{i\in\mathbb{N}} \cdot \{b_i\}_{i\in\mathbb{N}} = \{a_i b_i\}_{i\in\mathbb{N}}.$$

Then $\mathcal{C}$ together with these operations is a ring with the identity element. All null sequences $\{a_i\}_{i\in\mathbb{N}} \to 0$ form a maximal ideal and the corresponding quotient field $\mathbb{Q}_p$ is called the *field of p-adic numbers*. Note that this construction does not depend on the choice of $c \in (0,1)$. If σ is an endomorphism of $\mathbb{Q}_p$ then the restriction of σ to $\mathbb{Q}$ is identity and it follows directly from the definition of $\mathbb{Q}_p$ that σ is identity.

Example 1.6. Consider the division ring $\mathbb{H}$ formed by the real quaternion numbers

$$a + bi + cj + dk, \quad a,b,c,d \in \mathbb{R}.$$

This is the 4-dimensional real vector space $\mathbb{R}^4$ whose standard base is written as $1, i, j, k$. The multiplicative operation on $\mathbb{H}$ is defined by the following conditions

$$i^2 = j^2 = k^2 = -1$$

and

$$ij = k, \quad jk = i, \quad ki = j.$$

This division ring is non-commutative and the conjugate mapping

$$q = a + bi + cj + dk \to \overline{q} = a - bi - cj - dk$$

is an anti-automorphism. Every quaternion $a + bi + cj + dk$ can be presented as the pair consisting of the scalar a and the vector $(b, c, d) \in \mathbb{R}^3$. Denote by $\mathbb{H}_0$ the set of all quaternions whose scalar part is zero. Then $\mathbb{H} = \mathbb{R} + \mathbb{H}_0$. The following facts are obvious:

(1) the center of $\mathbb{H}$ coincides with $\mathbb{R}$;
(2) a quaternion q belongs to $\mathbb{H}_0$ if and only if its square q^2 is a non-positive real number; for every $q \in \mathbb{H}_0$ we have $q^2 = -|q|^2$, where $|q|$ is the absolute value of the vector $q \in \mathbb{R}^3 = \mathbb{H}_0$.

Let σ be an automorphism or an anti-automorphism of $\mathbb{H}$. Using (1) we establish that $\sigma(\mathbb{R}) = \mathbb{R}$. Thus the restriction of σ to $\mathbb{R}$ is identity. The property (2) guarantees that $\sigma(\mathbb{H}_0) = \mathbb{H}_0$ and σ induces an orthogonal linear automorphism of $\mathbb{H}_0 = \mathbb{R}^3$. So, there exists $u \in O(\mathbb{R}^3)$ such that for every quaternion $q = a + x$ with $a \in \mathbb{R}$ and $x \in \mathbb{H}_0$ we have

$$\sigma(q) = a + u(x).$$

Every reflection of the vector space $\mathbb{H}_0 = \mathbb{R}^3$ is induced by an anti-automorphism of type

$$q \to p\,\overline{q}\,p^{-1}, \quad p \neq 0 \tag{1.2}$$

and all rotations of this vector space are induced by inner automorphisms of $\mathbb{H}$, see [Porteous (1981), Chapter 10]. This implies that every automorphism of $\mathbb{H}$ is inner and every anti-automorphism is of type (1.2).

The statement concerning automorphisms of $\mathbb{H}$ is a partial case of the following remarkable result.

Theorem 1.1 (Skolem–Noether's theorem). *Let R be a division ring with center C. Suppose that R is a finite-dimensional vector space over C. If σ is an automorphism of R such that $\sigma|_C$ is identity then σ is an inner automorphism.*

Proof. See, for example, [Cohn (1995), Section 3.3]. $\qquad\qquad\square$

Example 1.7. By classical Wedderburn's theorem, every finite division ring is a field. Every finite field is isomorphic to one of the Galois fields. For a prime number $p > 1$ and an integer $n \geq 1$ the corresponding Galois field $\mathrm{GF}(p^n)$ is the subfield of the algebraic closure of $\mathbb{Z}_p$ formed by the roots of the polynomial equality

$$x^{p^n} - x = 0.$$

The field $\mathrm{GF}(p^n)$ consists of p^n elements and $\mathrm{GF}(p)$ coincides with $\mathbb{Z}_p$. The automorphism group of $\mathbb{Z}_p$ is trivial. If $n > 1$ then the automorphism group of $\mathrm{GF}(p^n)$ is generated by the Frobenius automorphism $a \to a^p$ whose order is equal to n. Every $\mathrm{GF}(p^n)$ is an extension of $\mathbb{Z}_p$ and $\mathrm{GF}(p^n)$ is an extension of $\mathrm{GF}(p^m)$ if and only if m divides n. In the case when $n = dm$, we have

$$[\mathrm{GF}(p^n) : \mathrm{GF}(p^m)] = d.$$

See [Lang (2002), Section VII.5] for the details.

Example 1.8. The well-known Frobenius theorem states that there are precisely three finite-dimensional associative division algebras over the field of real numbers — $\mathbb{R}, \mathbb{C}, \mathbb{H}$. The rational function field $\mathbb{R}(t)$ is infinite-dimensional over $\mathbb{R}$. Let σ be an endomorphism of $\mathbb{R}(t)$. Every positive real number a can be characterized as an element of $\mathbb{R}(t)$ such that for every natural n there is $b \in \mathbb{R}(t)$ satisfying $a = b^n$. This guarantees that

σ transfers $\mathbb{R}$ to itself. Then $\sigma|_{\mathbb{R}}$ is identity and σ is completely defined by the value in t. As in Example 1.3, σ is an automorphism if and only if

$$\sigma(t) = \frac{at + b}{ct + d},$$

where $a, b, c, d \in \mathbb{R}$ and $ad \neq bc$. Thus there exist non-surjective endomorphisms of $\mathbb{R}(t)$.

The previous example can be generalized as follows.

Example 1.9. Let F be a field. Denote by $F(t_1, \ldots, t_n)$ the field formed by all rational functions of n variables with coefficients in F. The automorphism group of this field is non-trivial and there are non-surjective endomorphisms.

We refer [Cohn (1995)] for various examples of non-commutative division rings.

Proposition 1.3 (Proposition 2.3.5 in [Cohn (1995)]). *For every field F there is a division ring R such that F is the center of R and R is an infinite-dimensional vector space over F.*

As above, we suppose that R and R' are division rings. Consider the following equivalence relation on the set of all homomorphisms of R to R'. Two homomorphisms $\sigma_1 : R \to R'$ and $\sigma_2 : R \to R'$ are *equivalent* if there exist automorphisms $\gamma \in \mathrm{Aut}(R)$ and $\gamma' \in \mathrm{Aut}(R')$ such that

$$\sigma_2 = \gamma' \sigma_1 \gamma.$$

If R and R' are isomorphic then any two isomorphisms of R to R' are equivalent and non-surjective homomorphisms of R to R' are not equivalent to isomorphisms of R to R'.

Example 1.10. Consider the case when R is a subring of R'. For every endomorphism $\sigma : R' \to R'$ the restriction to R is a homomorphism of R to R'. If σ is an automorphism of R' then $\sigma|_R$ is equivalent to the identity homomorphism of R to R'. Now we suppose that every automorphism of R can be extended to an automorphism of R'. This is true if R' is an algebraically closed field (Proposition 1.2); but this fails if $R' = \mathbb{R}$ and R is one of the fields considered in Examples 1.2, 1.3 or $R' = \mathbb{H}$ and $R = \mathbb{C}$. An easy verification shows that a homomorphism of R to R' is equivalent to the identity homomorphism of R to R' if and only if it is the restriction of an automorphism of R'.

1.2 Vector spaces over division rings

Let V be an abelian group whose identity element is denoted by 0. Suppose that R is a division ring and

$$R \times V \to V, \quad (a, x) \to ax$$

is a left action of this division ring on V. We say that V is a *left vector space* over R if the following conditions hold:

(L1) $1x = x$ for all $x \in V$,
(L2) $a(x + y) = ax + ay$ for all $a \in R$ and $x, y \in V$,
(L3) $(a + b)x = ax + bx$ for all $a, b \in R$ and $x \in V$,
(L4) $a(bx) = (ab)x$ for all $a, b \in R$ and $x \in V$.

The condition (L2) guarantees that $a0 = 0$ for every $a \in R$ and $0 \in V$. Similarly, (L3) implies that $0x = 0$ for every $x \in V$ and $0 \in R$. Elements of V and R are called *vectors* and *scalars*, respectively.

Every right action

$$V \times R \to V$$

defines a *right vector space* over R if it satisfies the following right versions of the conditions (L1)–(L4):

(R1) $x1 = x$ for all $x \in V$,
(R2) $(x + y)a = xa + ya$ for all $a \in R$ and $x, y \in V$,
(R3) $x(a + b) = xa + xb$ for all $a, b \in R$ and $x \in V$,
(R4) $(xb)a = x(ba)$ for all $a, b \in R$ and $x \in V$.

We can consider any right action as a left action and rewrite (R4) as follows

$$a(bx) = (ba)x = (a * b)x,$$

where $*$ is the multiplicative operation of the opposite division ring. Therefore, every right vector space over R is a left vector space over R^*. For this reason, we do not distinguish left and right vector spaces over fields.

Remark 1.2. The multiplicative operation $R \times R \to R$ defines left and right actions on R satisfying (L1)–(L4) and (R1)–(R4), respectively. Thus every division ring can be considered as a left vector space and a right vector space over itself.

Let V be a left vector space over a division ring R. A subset $S \subset V$ is said to be a *subspace* of V if for any scalars $a, b \in R$ and any vectors

$x, y \in S$ the linear combination $ax + by$ belongs to S. It is clear that $\{0\}$ and V are subspaces and the intersection of any collection of subspaces is a subspace. Every subspace of V can be considered as a left vector space over R.

The minimal subspace containing a subset $X \subset V$ (the intersection of all subspaces containing X) is called *spanned* by X and denoted by $\langle X \rangle$. This subspace is formed by all linear combinations of vectors from X. We say that X is an *independent* subset if the subspace $\langle X \rangle$ cannot be spanned by a proper subset of X, or equivalently, any distinct vectors $x_1, \ldots, x_k \in X$ are *linearly independent*, i.e. the equality

$$a_1 x_1 + \cdots + a_k x_k = 0$$

holds only in the case when every scalar a_i is zero. Non-zero vectors $x_1, \ldots, x_k$ are *linearly dependent* if in the latter equality some of a_i are non-zero.

Let S be a subspace of V (possibly $S = V$). An independent subset $B \subset S$ is a *base* of S if $\langle B \rangle$ coincides with S. All bases of S have the same cardinality (possibly infinite) called the *dimension* of S and denoted by $\dim S$. Every base of S can be extended to a base of V.

Example 1.11. The Cartesian n-product

$$R^n = \underbrace{R \times \cdots \times R}_{n}$$

of our division ring together with the operations

$$(a_1, \ldots, a_n) + (b_1, \ldots, b_n) := (a_1 + b_1, \ldots, a_n + b_n)$$

and

$$a(a_1, \ldots, a_n) := (aa_1, \ldots, aa_n)$$

is an n dimensional left vector space over R. If the multiplicative operation is replaced by

$$(a_1, \ldots, a_n)a := (a_1 a, \ldots, a_n a)$$

then we get an n-dimensional right vector space over R.

For subsets $X, Y \subset V$ we define the *sum* $X + Y$ as the set of all vectors $x + y$, where $x \in X$ and $y \in Y$. If X, Y are subspaces then $X + Y$ coincides with $\langle X, Y \rangle$ (the subspace spanned by X and Y) and

$$\dim(X + Y) = \dim X + \dim Y - \dim(X \cap Y).$$

Lemma 1.1. *For any subspaces $S, U \subset V$ there is a base of V such that S and U are spanned by subsets of this base.*

Proof. We take any base of $S \cap U$ and extend it to bases of S and U. Let B be the union of these bases. Then $\langle B \rangle = S + U$. The equality

$$\dim(S + U) = \dim S + \dim U - \dim(S \cap U) = |B|$$

shows that B is a base of $S + U$. Any base of V containing B is as required.

$\square$

For a subspace $S \subset V$ the associated *quotient vector space* V/S is formed by subsets of type $x + S$ (we have $x + S = y + S$ if and only if $x - y \in S$). The vector operations on V/S are defined by the formulas

$$(x + S) + (y + S) := (x + y) + S,$$

$$a(x + S) := ax + S.$$

The dimension of V/S is equal to $\dim V - \dim S$.

Suppose that $\dim V = n$ is finite and write $\mathcal{G}(V)$ for the set of all subspaces of V. For every $k \in \{0, 1, \ldots, n\}$ we denote by $\mathcal{G}_k(V)$ the Grassmannian consisting of all k-dimensional subspaces of V. Then $\mathcal{G}_0(V) = \{0\}$ and $\mathcal{G}_n(V) = \{V\}$.

Now we suppose that $k \in \{1, \ldots, n - 1\}$ and S, U are subspaces of V such that

$$S \subset U \quad \text{and} \quad \dim S < k < \dim U.$$

We define

$$[S, U]_k := \{\, P \in \mathcal{G}_k(V) \ : \ S \subset P \subset U \,\}.$$

In the case when $S = 0$ or $U = V$, this set will be denoted by $\langle U]_k$ or $[S\rangle_k$, respectively. It is clear that $[S, U]_k$ can be naturally identified with the Grassmannian $\mathcal{G}_{k-s}(U/S)$, where $s = \dim S$.

1.3 Semilinear mappings

Let V and V' be left vector spaces over division rings R and R', respectively. Following [Faure and Frölicher (2000), 6.3.1] a mapping $l : V \to V'$ is called *semilinear* if

$$l(x + y) = l(x) + l(y)$$

for all $x, y \in V$ (the mapping l is additive) and there exists a homomorphism $\sigma : R \to R'$ such that

$$l(ax) = \sigma(a)l(x)$$

for all $a \in R$ and $x \in V$. If l is non-zero then there is the unique homomorphism $\sigma : R \to R'$ satisfying this condition and we say that l is σ-*linear*. In the case when $R = R'$ and σ is identity, the mapping l is *linear*. Linear and semilinear mappings of V to itself are said to be linear and semilinear *endomorphisms* of V.

Example 1.12. If $l : V \to V'$ is a σ-linear mapping and $a \in R' \setminus \{0\}$ then the scalar multiple al (the mapping transferring every x to $al(x)$) is γ-linear, where $\gamma(b) = a\sigma(b)a^{-1}$ for every $b \in R$.

Example 1.13. Suppose that $\dim V = 1$. Then V can be identified with R. Every linear endomorphism of V is of type $x \to xa$. Every semilinear endomorphism of V is of type $x \to \sigma(x)a$, where σ is an endomorphism of R.

A semilinear mapping $l : V \to V'$ transfers subspaces of V to subspaces of V' only in the case when the associated homomorphism of R to R' is an isomorphism. In the general case, we have the following.

Lemma 1.2 *If $l : V \to V'$ is a semilinear mapping then*

$$\langle l(X \cup Y) \rangle = \langle l(X) \rangle + \langle l(Y) \rangle$$

for any subsets $X, Y \subset V$.

Proof. Easy verification. □

A semilinear mapping of V to V' is called a *semilinear isomorphism* if it is bijective and the associated homomorphism of R to R' is an isomorphism. For every σ-linear isomorphism of V to V' the inverse mapping is a (σ^{-1})-linear isomorphism of V' to V.

The following example shows that there exist semilinear bijections over non-surjective homomorphisms of division rings.

Example 1.14. The complexification mapping of $\mathbb{R}^{2n}$ to $\mathbb{C}^n$

$$(x_1, y_1, \ldots, x_n, y_n) \to (x_1 + y_1 i, \ldots, x_n + y_n i)$$

is a semilinear bijection over the identity homomorphism of $\mathbb{R}$ to $\mathbb{C}$. Similarly, the mapping of $\mathbb{R}^{4n}$ to $\mathbb{H}^n$ transferring every vector

$$(a_1, b_1, c_1, d_1, \ldots, a_n, b_n, c_n, d_n)$$

to the vector

$$(a_1 + b_1 i + c_1 j + d_1 k, \ldots, a_n + b_n i + c_n j + d_n k)$$

is a semilinear bijection over the identity homomorphism of $\mathbb{R}$ to $\mathbb{H}$.

Semilinear isomorphisms transfer independent subsets to independent subsets and bases to bases. Therefore, the existence of semilinear isomorphisms of V to V' implies that $\dim V = \dim V'$ and R is isomorphic to R'. In this case, for any isomorphism $\sigma : R \to R'$ and any bases $\{x_i\}_{i \in I}$ and $\{x_i'\}_{i \in I}$ of V and V' (respectively) there is the unique σ-linear isomorphism of V to V' sending every x_i to x_i'.

The group of all semilinear automorphisms of V (semilinear isomorphisms of V to itself) is denoted by $\Gamma L(V)$. The general linear group $GL(V)$ formed by all linear automorphisms of V is a normal subgroup of $\Gamma L(V)$ and the corresponding quotient group is isomorphic to $\mathrm{Aut}(R)$.

Example 1.15. Let $a \in R \setminus \{0\}$. By Example 1.12, the associated homothety $x \to ax$ is a semilinear automorphism of V. It is linear if and only if a belongs to the center of the division ring R. The group of all homothetic transformations $H(V)$ is a normal subgroup of $\Gamma L(V)$ isomorphic to the multiplicative group of the division ring. Denote by $H_0(V)$ the group of all linear homothetic transformations. This is a normal subgroup of $GL(V)$ isomorphic to the center of R.

Lemma 1.3. *Let* $l : V \to V'$ *be a* σ-*linear mapping. Let also* S *and* S' *be subspaces of* V *and* V' *(respectively) such that* $l(S) \subset S'$. *Then* l *induces a* σ-*linear mapping*

$$\tilde{l} : V/S \to V'/S'. \tag{1.3}$$

If $l(S) = S'$ *and* l *is a semilinear isomorphism then* $\tilde{l}$ *is a semilinear isomorphism.*

Proof. Easy verification. ⊔

Remark 1.3. Conversely, consider a σ-linear mapping (1.3). Let W and W' be complements of S and S' in V and V', respectively. For every $x \in W$ there is unique $x' \in W'$ (possibly zero) such that

$$\tilde{l}(x + S) = x' + S'.$$

An easy verification shows that the correspondence $x \to x'$ is a σ-linear mapping of W to W'; moreover, this is a semilinear isomorphism if $\tilde{l}$ is a semilinear isomorphism. We take any σ-linear mapping of S to S' and extend our σ-linear mapping of W to W' to a σ-linear mapping of V to V'. By Lemma 1.3, every such an extension (it is not unique) induces a σ-linear mapping of V/S to V'/S'. The latter mapping always coincides with $\tilde{l}$.

Example 1.16. Suppose that R is non-commutative and C is the center of R. Then V can be considered as a vector space over the field C. We denote this vector space by V_C. Let σ be an automorphism of R. Then $\sigma(C) = C$ and every σ-linear endomorphism of V is a $(\sigma|_C)$-linear endomorphism of V_C. In the case when R is a finite-dimensional vector space over C, this semilinear endomorphism of V_C is linear if and only if the automorphism σ is inner (Theorem 1.1). In particular, all semilinear mappings between quaternion vector spaces associated to automorphisms of $\mathbb{H}$ are linear mappings between the corresponding real vector spaces.

There is the following equivalence relation on the set of all semilinear mappings of V to V'. Semilinear mappings $l_1 : V \to V'$ and $l_2 : V \to V'$ are *equivalent* if

$$l_2 = u'l_1 u$$

for some semilinear automorphisms $u \in \Gamma L(V)$ and $u' \in \Gamma L(V')$. All semilinear isomorphisms of V to V' (if they exist) are equivalent. For equivalent semilinear mappings of V to V' the associated homomorphisms of R to R' are equivalent.

A semilinear mapping $l : V \to V'$ is injective if and only if the kernel $\mathrm{Ker}(l)$ (the subspace formed by all vectors $x \in V$ satisfying $l(x) = 0$) is zero. If $l : V \to V'$ is a semilinear injection over an isomorphism of R to R' then $l(V)$ is a subspace of V' and l is a semilinear isomorphism of V to $l(V)$. Semilinear injections over non-surjective homomorphisms of division rings can be more complicated. For example, if $l : V \to V'$ is one of the semilinear bijections considered in Example 1.14 then there are distinct 1-dimensional subspaces $P_1, P_2 \subset V$ whose images are contained in the same 1-dimensional subspace of V' and we have $\langle l(P_1) \rangle = \langle l(P_2) \rangle$.

1.4   Semilinear embeddings

As in the previous section, we consider left vector spaces V and V' over division rings R and R', respectively. We suppose that $\dim V = n$ and $\dim V' = n'$ both are finite. Let $m \in \{2, \ldots, n\}$. A semilinear injection $l : V \to V'$ is said to be a *semilinear m-embedding* if it transfers any m linearly independent vectors to linearly independent vectors. In the case when $m = n$, we say that l is a *strong semilinear embedding*.

Strong semilinear embeddings send independent subsets to independent subsets. If a semilinear injection transfers at least one base of the vector space to an independent subset then it is a strong semilinear embedding. The existence of strong semilinear embeddings of V in V' implies that $n \leq n'$. In this case, for any homomorphism $\sigma : R \to R'$, any base $x_1, \ldots, x_n$ of V and any linearly independent vectors $x'_1, \ldots, x'_n \in V'$ there is the unique strong σ-linear embedding of V in V' sending every x_i to x'_i.

Proposition 1.4. *Strong semilinear embeddings of V in V' (if they exist) are equivalent if and only if the associated homomorphisms of R to R' are equivalent.*

Proof. Let l_1, l_2 be strong semilinear embeddings of V in V' and let σ_1, σ_2 be the associated homomorphisms of R to R'. The equivalence of l_1 and l_2 implies the equivalence of σ_1 and σ_2. Conversely, suppose that σ_1 and σ_2 are equivalent, i.e. there exist automorphisms $\gamma \in \mathrm{Aut}(R)$ and $\gamma' \in \mathrm{Aut}(R')$ such that

$$\sigma_2 = \gamma' \sigma_1 \gamma. \tag{1.4}$$

Let $x_1, \ldots, x_n$ be a base of V. Then

$$\{l_1(x_1), \ldots, l_1(x_n)\} \quad \text{and} \quad \{l_2(x_1), \ldots, l_2(x_n)\}$$

are independent subsets of V'. Consider the γ-linear automorphism u of V leaving fixed every x_i and any γ'-linear automorphism u' of V' sending every $l_1(x_i)$ to $l_2(x_i)$. The equality (1.4) shows that $u'l_1u$ is a strong σ_2-linear embedding of V in V'. Since it transfers every x_i to $l_2(x_i)$, we have $l_2 = u'l_1u$. $\qquad\square$

Now we construct a semilinear m-embedding of an $(m+1)$-dimensional vector space in an m-dimensional vector space. Our example is a modification of the example given in [Brezulcanu and Rădulescu (1984)].

Example 1.17. Let F and E be fields and let σ be a homomorphism of F to E. Suppose that

$$[E : \sigma(F)] \geq 4.$$

Then there exist $t_1, t_2, t_3 \in E$ such that $1, t_1, t_2, t_3$ are linearly independent over $\sigma(F)$. Consider the σ-linear mapping $l : F^4 \to E^3$ defined as follows

$$l(a_1, a_2, a_3, a_4) = (\sigma(a_1) + \sigma(a_4)t_1, \sigma(a_2) + \sigma(a_4)t_2, \sigma(a_3) + \sigma(a_4)t_3). \tag{1.5}$$

This mapping is injective (since $1, t_1, t_2, t_3$ are linearly independent over $\sigma(F)$). Show that l is a semilinear 3-embedding. Let

$$x_i = (a_{i1}, a_{i2}, a_{i3}, a_{i4}), \quad i \in \{1, 2, 3\}$$

be linearly independent vectors of F^4. Consider the matrix

$$M = \begin{pmatrix} \sigma(a_{11}) + \sigma(a_{14})t_1 & \sigma(a_{12}) + \sigma(a_{14})t_2 & \sigma(a_{13}) + \sigma(a_{14})t_3 \\ \sigma(a_{21}) + \sigma(a_{24})t_1 & \sigma(a_{22}) + \sigma(a_{24})t_2 & \sigma(a_{23}) + \sigma(a_{24})t_3 \\ \sigma(a_{31}) + \sigma(a_{34})t_1 & \sigma(a_{32}) + \sigma(a_{34})t_2 & \sigma(a_{33}) + \sigma(a_{34})t_3 \end{pmatrix}$$

whose rows are $l(x_1), l(x_3), l(x_3)$. The determinant of M is equal to

$$\sigma\left(\begin{vmatrix} a_{11} & a_{12} & a_{13} \\ a_{21} & a_{22} & a_{23} \\ a_{31} & a_{32} & a_{33} \end{vmatrix} \right) + \sigma\left(\begin{vmatrix} a_{14} & a_{12} & a_{13} \\ a_{24} & a_{22} & a_{23} \\ a_{34} & a_{32} & a_{33} \end{vmatrix} \right) t_1$$

$$+ \sigma\left(\begin{vmatrix} a_{11} & a_{14} & a_{13} \\ a_{21} & a_{24} & a_{23} \\ a_{31} & a_{34} & a_{33} \end{vmatrix} \right) t_2 + \sigma\left(\begin{vmatrix} a_{11} & a_{12} & a_{14} \\ a_{21} & a_{22} & a_{24} \\ a_{31} & a_{32} & a_{34} \end{vmatrix} \right) t_3.$$

Since $1, t_1, t_2, t_3$ are linearly independent over $\sigma(F)$ and σ is injective, we have $\det(M) = 0$ only in the case when

$$\begin{vmatrix} a_{11} & a_{12} & a_{13} \\ a_{21} & a_{22} & a_{23} \\ a_{31} & a_{32} & a_{33} \end{vmatrix} = \begin{vmatrix} a_{14} & a_{12} & a_{13} \\ a_{24} & a_{22} & a_{23} \\ a_{34} & a_{32} & a_{33} \end{vmatrix} = \begin{vmatrix} a_{11} & a_{14} & a_{13} \\ a_{21} & a_{24} & a_{23} \\ a_{31} & a_{34} & a_{33} \end{vmatrix} = \begin{vmatrix} a_{11} & a_{12} & a_{14} \\ a_{21} & a_{22} & a_{24} \\ a_{31} & a_{32} & a_{34} \end{vmatrix} = 0.$$

The latter contradicts the fact that x_1, x_2, x_3 are linearly independent. Therefore, $\det(M) \neq 0$ which means that $l(x_1), l(x_3), l(x_3)$ are linearly independent. Similarly, we construct a σ-linear m-embedding of F^{m+1} in E^m if

$$[E : \sigma(F)] \geq m + 1.$$

Now we suppose that $E = F(t)$ is the field of rational functions with coefficients in F. Then every endomorphism σ of F can be considered as a homomorphism of F to E. Since $[E : \sigma(F)]$ is infinite, for every integer m there is a σ-linear m-embedding of F^{m+1} in E^m.

Remark 1.4. As in Example 1.17, we suppose that $l : F^4 \to E^3$ is the σ-linear mapping defined by (1.5); but now we consider the case when $1, t_1, t_2, t_3$ are linearly dependent over $\sigma(F)$, i.e.

$$a_0 + a_1 t_1 + a_2 t_2 + a_3 t_3 = 0,$$

where all a_i belong to $\sigma(F)$ and some of them are non-zero. We want to show that there exist vectors

$$x_i = (a_{i1}, a_{i2}, a_{i3}, a_{i4}), \quad i \in \{1, 2, 3\}$$

in F^4 such that

$$\begin{vmatrix} a_{11} & a_{12} & a_{13} \\ a_{21} & a_{22} & a_{23} \\ a_{31} & a_{32} & a_{33} \end{vmatrix} = \sigma^{-1}(a_0), \quad \begin{vmatrix} a_{14} & a_{12} & a_{13} \\ a_{24} & a_{22} & a_{23} \\ a_{34} & a_{32} & a_{33} \end{vmatrix} = \sigma^{-1}(a_1),$$

$$\begin{vmatrix} a_{11} & a_{14} & a_{13} \\ a_{21} & a_{24} & a_{23} \\ a_{31} & a_{34} & a_{33} \end{vmatrix} = \sigma^{-1}(a_2), \quad \begin{vmatrix} a_{11} & a_{12} & a_{14} \\ a_{21} & a_{22} & a_{24} \\ a_{31} & a_{32} & a_{34} \end{vmatrix} = \sigma^{-1}(a_3).$$

Since some of a_i are non-zero, these vectors are linearly independent. If $a_0 \neq 0$ then we put

$$a_{31} = a_{32} = a_{13} = a_{23} = 0$$

and choose $a_{11}, a_{12}, a_{21}, a_{22}, a_{33}$ satisfying

$$\begin{vmatrix} a_{11} & a_{12} & 0 \\ a_{21} & a_{22} & 0 \\ 0 & 0 & a_{33} \end{vmatrix} = \sigma^{-1}(a_0) \neq 0.$$

The system of linear equations

$$\begin{vmatrix} x & a_{12} & 0 \\ y & a_{22} & 0 \\ z & 0 & a_{33} \end{vmatrix} = \sigma^{-1}(a_1), \quad \begin{vmatrix} a_{11} & x & 0 \\ a_{21} & y & 0 \\ 0 & z & a_{33} \end{vmatrix} = \sigma^{-1}(a_2), \quad \begin{vmatrix} a_{11} & a_{12} & x \\ a_{21} & a_{22} & y \\ 0 & 0 & z \end{vmatrix} = \sigma^{-1}(a_3)$$

is non-degenerate and we take

$$a_{14} = x, \ a_{24} = y, \ a_{34} = z,$$

where x, y, z is the solution of this system. In the case when $a_0 - 0$, we apply similar arguments to any $a_i \neq 0$. So, we have three vectors x_1, x_2, x_3 satisfying the required conditions. If M is the matrix whose rows are $l(x_1), l(x_3), l(x_3)$ then

$$\det(M) = a_0 + a_1 t_1 + a_2 t_2 + a_3 t_3 = 0$$

and $l(x_1), l(x_3), l(x_3)$ are linearly dependent. Thus l is not a semilinear 3-embedding.

Using the arguments from Example 1.17 and Remark 1.4 we can prove the following.

Proposition 1.5. *Suppose that R and R' both are fields and $l : V \to V'$ is a σ-linear mapping. Then the following assertions are fulfilled:*

(1) *If there exists a base $x_1, \ldots, x_n$ of V such that $l(x_1), \ldots, l(x_{n-1})$ are linearly independent in V' and*

$$l(x_n) = t_1 l(x_1) + \cdots + t_{n-1} l(x_{n-1}),$$

where each scalar t_i does not belong to $\sigma(R)$ and $1, t_1, \ldots, t_{n-1}$ are linearly independent over $\sigma(R)$ then l is a semilinear $(n-1)$-embedding.

(2) *If l is a semilinear $(n - 1)$-embedding then every base $x_1, \ldots, x_n$ of V satisfies the above conditions.*

Remark 1.5. The determinant function is defined for matrices over noncommutative division rings [Artin (1957), Section IV.1]; but in contrast to the commutative case, it is not multilinear on rows and columns. For this reason, we are not be able prove the same statements in the general case.

We use Example 1.17 to construct a semilinear m-embedding of an $(m + k)$-dimensional vector space in an m-dimension vector space. Other example can be found in [Kreuzer 1 (1998)].

Example 1.18. Consider a sequence of field extensions

$$E_1 \subset E_2 \subset \cdots \subset E_{k+1}$$

such that

$$[E_{i+1} : E_i] \geq m + k - i + 1$$

for every $i \in \{1, \ldots, k\}$. For example, we can suppose that E_{i+1} is the field of rational functions with coefficients in E_i. Also, for every prime number $p > 1$ we can choose integers $n_1, \ldots, n_{k+1}$ such that

$$E_1 = \mathrm{GF}(p^{n_1}), \ldots, E_{k+1} = \mathrm{GF}(p^{n_{k+1}})$$

are as required, see Example 1.7. For every $i \in \{1, \ldots, k + 1\}$ we take a vector space W_i over E_i such that

$$\dim W_i = m + k - i + 1.$$

By Example 1.17, there is a semilinear $(m + k - i)$-embedding

$$l_i : W_i \to W_{i+1}.$$

The composition $l_k l_{k-1} \ldots l_1$ is a semilinear m-embedding of the $(m + k)$-dimensional vector space W_1 in the m-dimensional vector space W_{k+1}.

By Proposition 1.4, if $n \le n'$ then there is a one-to-one correspondence between the equivalence classes of strong semilinear embeddings of V in V' and the equivalence classes of homomorphisms of R to R'. The following example shows that the case of non-strong semilinear embeddings is more complicated.

Example 1.19. Suppose that $E = \mathbb{R}(t)$ is the rational function field over $\mathbb{R}$. Consider the semilinear n-embeddings l and s of $\mathbb{R}^{n+1}$ in E^n defined as follows

$$l(a_1, \ldots, a_{n+1}) = (a_1 + a_{n+1}t, \ldots, a_n + a_{n+1}t^n),$$

$$s(a_1, \ldots, a_{n+1}) = (a_1 + a_{n+1}t^{n+1}, \ldots, a_n + a_{n+1}t^{2n}),$$

see Example 1.17. By Example 1.8, for every automorphism $\sigma \in \mathrm{Aut}(E)$ we have

$$\sigma(t^m) = \left(\frac{at + b}{ct + d}\right)^m \quad \forall\, m \in \mathbb{Z},$$

where $a, b, c, d \in \mathbb{R}$ and $ad \ne bc$. Then

$$\sigma(t) = \frac{at + b}{ct + d}, \ldots, \sigma(t^n) = \left(\frac{at + b}{ct + d}\right)^n, t^{n+1}, \ldots, t^{2n}$$

are linearly independent over $\mathbb{R}$. Since the restriction of every automorphism of E to $\mathbb{R}$ is identity, l and s cannot be equivalent. The same arguments show that there are infinitely many equivalence classes of semilinear n-embeddings of $\mathbb{R}^{n+1}$ in E^n over the identity homomorphism of $\mathbb{R}$ to E.

1.5 Mappings of Grassmannians induced by semilinear embeddings

Recall that V and V' are left vector spaces over division rings R and R' (respectively) and $\dim V = n$, $\dim V' = n'$ both are finite. Every semilinear injection $l : V \to V'$ induces the mapping

$$(l)_1 : \mathcal{G}_1(V) \to \mathcal{G}_1(V')$$

transferring every 1-dimensional subspace $P \subset V$ to the 1-dimensional subspace of V' containing $l(P)$. This mapping is not necessarily injective (for example, it is non-injective if l is one of the mappings considered in Example 1.14). For every non-zero scalar $a \in R'$ we have $(l)_1 = (al)_1$.

Proposition 1.6. *Let l and s be semilinear injections of V to V' such that $(l)_1$ coincides with $(s)_1$. If $l(V)$ contains two linearly independent vectors then s is a scalar multiple of l.*

Proof. The equality $(l)_1 = (s)_1$ implies that for every non-zero vector $x \in V$ there is a non-zero scalar $a_x \in R'$ such that

$$s(x) = a_x l(x).$$

Let $x, y \in V \setminus \{0\}$. If $l(x)$ and $l(y)$ are linearly independent then we have

$$a_{x+y} l(x) + a_{x+y} l(y) = a_{x+y} l(x+y)$$

$$= s(x+y) = s(x) + s(y) = a_x l(x) + a_y l(y)$$

which implies that

$$a_x = a_{x+y} = a_y.$$

In the case when $l(y)$ is a scalar multiple of $l(x)$, we take $z \in V$ such that $l(x)$ and $l(z)$ are linearly independent (this is possible, since $l(V)$ contains two linearly independent vectors). Using the above arguments, we show that $a_x = a_z = a_y$. $\qquad\square$

Let $m \in \{2, \ldots, n\}$ and let $l : V \to V'$ be a semilinear m-embedding. If S is a subspace of V whose dimension is not greater than m then S and $\langle l(S) \rangle$ are of the same dimension. Therefore, for every $k \in \{1, \ldots, m\}$ we have the mapping

$$(l)_k : \mathcal{G}_k(V) \to \mathcal{G}_k(V')$$

$$S \to \langle l(S) \rangle.$$

If l is a strong semilinear embedding then the mapping $(l)_k$ is defined for every $k \in \{1, \ldots, n-1\}$ (we do not consider the trivial case when $k = n$).

Proposition 1.7. *The following assertions are fulfilled:*

(1) *the mapping $(l)_k$ is injective for every $k \in \{1, \ldots, m-1\}$,*
(2) *if $m < n$ and $(l)_m$ is injective then l is a semilinear $(m+1)$-embedding.*

Proof. (1). If $(l)_k$ is non-injective for a certain $k \in \{1, \ldots, m-1\}$ then there exist distinct k-dimensional subspaces $S, U \subset V$ such that $l(S)$ and $l(U)$ span the same k-dimensional subspace of V'. By Lemma 1.2, this k-dimensional subspace coincides with $\langle l(S + U) \rangle$. Hence it contains the image of every $(k+1)$-dimensional subspace $T \subset S + U$ (the dimension of $S + U$ is not less than $k+1$, since S and U are distinct). On the other hand, if $T \in \mathcal{G}_{k+1}(V)$ then $l(T)$ spans a $(k+1)$-dimensional subspace of V' (since $k+1 \leq m$) and we get a contradiction.

(2). Suppose that the mapping $(l)_m$ is injective and there is an independent subset $X \subset V$ consisting of $m+1$ vectors and such that $l(X)$ is not independent. The dimension of $\langle l(X) \rangle$ is equal to m. If S and U are distinct m-dimensional subspaces spanned by subsets of X then $\langle l(S) \rangle$ and $\langle l(U) \rangle$ both coincide with $\langle l(X) \rangle$ which contradicts the injectivity of $(l)_m$. $\qquad\square$

If l is a semilinear isomorphism then $n = n'$ and $(l)_k$ is bijective for every $k \in \{1, \ldots, n-1\}$.

Proposition 1.8. *Suppose that $n \geq 3$, the mapping $(l)_k$ is bijective for a certain $k \in \{1, \ldots, m-1\}$ and one of the following possibilities is realized:*

(1) $k \leq m-2$,
(2) $n \leq n'$.

Then l is a semilinear isomorphism.

Proof. Let $\sigma : R \to R'$ be the homomorphism associated to l. It is sufficient to show that σ is surjective. The latter means that σ is an isomorphism of R to R' and $l(V)$ is a subspace of V'. Then $l(V) = V'$ (since $(l)_k$ is bijective) and we get the claim.

First we consider the case when $k = 1$.

The condition (1) implies that $m \geq 3$ and l transfers every triple of linearly independent vectors to a triple linearly independent vectors. We take any linearly independent vectors $x, y \in V \setminus \{0\}$. If $z \in V$ does not belong to $\langle x, y \rangle$ then $l(z)$ does not belong to $\langle l(x), l(y) \rangle$. This means that the restriction of $(l)_1$ to

$$\mathcal{G}_1(\langle x, y \rangle) = \{\langle x + ay \rangle : a \in R\} \cup \{\langle y \rangle\}$$

is a bijection to

$$\mathcal{G}_1(\langle l(x), l(y) \rangle) = \{\langle l(x) + a'l(y) \rangle : a' \in R'\} \cup \{\langle l(y) \rangle\}.$$

Since $(l)_1$ transfers $\langle x + ay \rangle$ to

$$\langle l(x) + \sigma(a)l(y) \rangle,$$

the homomorphism σ must be surjective.

Let $n \leq n'$, i.e. the condition (2) holds. Suppose that there exist linearly independent vectors $x, y, z \in V \setminus \{0\}$ such that $l(x), l(y), l(z)$ are linearly dependent. Let B be a base of V containing x, y, z. The subset $l(B)$ is not independent and the dimension of $\langle l(V) \rangle = \langle l(B) \rangle$ is less than

$n \leq n'$ which implies that $\langle l(V) \rangle$ is a proper subspace of V'. The latter is impossible, since $(l)_1$ is bijective. Therefore, l sends every triple of linearly independent vectors to a triple linearly independent vectors. As above, we show that σ is surjective.

Now we suppose that $k > 1$. For every $S \in \mathcal{G}_{k-1}(V)$ the mapping $(l)_k$ transfers $[S\rangle_k$ to a subset of $[S_l\rangle_k$, where

$$S_l := \langle l(S) \rangle = (l)_{k-1}(S) \in \mathcal{G}_{k-1}(V').$$

By the first part of Proposition 1.7, the mapping $(l)_{k-1}$ is injective and for any distinct $S, U \in \mathcal{G}_{k-1}(V)$ we have $S_l \neq U_l$. Since $(l)_k$ is bijective, the image of $[S\rangle_k$ coincides with $[S_l\rangle_k$ for every $S \in \mathcal{G}_{k-1}(V)$. Let us fix $S \in \mathcal{G}_{k-1}(V)$ and consider the σ-linear mapping

$$s : V/S \to V'/S_l$$

induced by l (Lemma 1.3). This is a semilinear $(m - k + 1)$-embedding and $(s)_1$ is a bijection of $\mathcal{G}_1(V/S)$ to $\mathcal{G}_1(V'/S_l)$. It follows from our assumption that one of the following possibilities is realized:

(1') $m - k + 1 \geq 3$, i.e. s sends any triple of linearly independent vectors to a triple of linearly independent vectors;

(2') $\dim V/S \leq \dim V'/S_l$.

We apply to $(s)_1$ the above arguments and establish that σ is surjective. □

In the next section we show that there is a semilinear 3-embedding of a vector space whose dimension is greater than 3 (and possibly infinite) in a 3-dimensional vector space which induces a bijection between the Grassmannians of 2-dimensional subspaces. The following problem is still open.

Problem 1.1. Suppose that $n > n'$ and $k = m - 1$, i.e. the conditions (1) and (2) from Proposition 1.8 do not hold. Is there a semilinear m-embedding $l : V \to V'$ such that $(l)_k$ is bijective?

Corollary 1.1. *Suppose that l is a strong semilinear embedding. Then $(l)_k$ is injective for every $k \in \{1, \ldots, n-1\}$. If $(l)_k$ is bijective for a certain integer $k \in \{1, \ldots, n-1\}$ then l is a semilinear isomorphism.*

Proof. The statement follows from Propositions 1.7 and 1.8. □

Corollary 1.2. *Let F and E be division rings. Suppose that F is a proper subring of E. Then for every natural $m \geq 2$ and every $k \in \{1, \ldots, m-1\}$ there exists $U \in \mathcal{G}_k(E^m)$ such that the dimension of the subspace $U \cap F^m$ in the vector space F^m is less than k.*

Proof. By Corollary 1.1, the mapping of $\mathcal{G}_k(F^m)$ to $\mathcal{G}_k(E^m)$ induced by the identity embedding of F^m in E^m is non-surjective. $\qquad\square$

If $a \in R'$ is non-zero then $(al)_k$ coincides with $(l)_k$ for all admissible k.

Proposition 1.9. *If $s : V \to V'$ is a semilinear m-embedding such that $(s)_k$ coincides with $(l)_k$ for a certain $k \in \{1, \ldots, m-1\}$ then s is a scalar multiple of l.*

Proof. In the case when $k = 1$, the statement follows from Proposition 1.6. Suppose that $k > 1$. For every $S \in \mathcal{G}_{k-1}(V)$ there exist

$$U_1, U_2 \in \mathcal{G}_k(V) \quad \text{such that} \quad S = U_1 \cap U_2.$$

By the first part of Proposition 1.7, the mapping $(l)_k$ is injective. Hence $\langle l(U_1)\rangle$ and $\langle l(U_2)\rangle$ are distinct elements of $\mathcal{G}_k(V')$ and

$$\dim(\langle l(U_1)\rangle \cap \langle l(U_2)\rangle) \le k - 1.$$

Since $\langle l(U_1)\rangle \cap \langle l(U_2)\rangle$ contains the $(k-1)$-dimensional subspace $\langle l(S)\rangle$, we have

$$\langle l(S)\rangle = \langle l(U_1)\rangle \cap \langle l(U_2)\rangle.$$

Similarly, we show that

$$\langle s(S)\rangle = \langle s(U_1)\rangle \cap \langle s(U_2)\rangle.$$

By our hypothesis,

$$\langle l(U_i)\rangle = \langle s(U_i)\rangle, \quad i = 1, 2$$

which implies that $\langle l(S)\rangle$ coincides with $\langle s(S)\rangle$. Thus $(l)_{k-1} = (s)_{k-1}$. Step by step, we establish that $(l)_1 = (s)_1$ which gives the claim. $\qquad\square$

It follows from Proposition 1.9 that the group of all transformations of $\mathcal{G}_k(V)$ induced by semilinear automorphisms of V can be naturally identified with the projective semilinear group

$$\mathrm{P\Gamma L}(V) := \Gamma \mathrm{L}(V)/\mathrm{H}(V).$$

Similarly, the group of all transformations of $\mathcal{G}_k(V)$ induced by linear automorphisms of V will be identified with the projective linear group

$$\mathrm{PGL}(V) := \mathrm{GL}(V)/\mathrm{H}_0(V).$$

Remark 1.6. Suppose that $u \in \Gamma \mathrm{L}(V)$ induces an element of $\mathrm{PGL}(V)$. By Proposition 1.9, u is a scalar multiple of a linear automorphism. Then one of the following possibilities is realized:

- R is a field and u is linear,
- R is non-commutative and the automorphism associated to u is inner.

Therefore, if R is a field then $\mathrm{P\Gamma L}(V)$ coincides with $\mathrm{PGL}(V)$ only in the case when the automorphism group of R is trivial. In the case when R is non-commutative, $\mathrm{P\Gamma L}(V)$ coincides with $\mathrm{PGL}(V)$ if and only if every automorphism of R is inner. The latter holds, for example, for the quaternion division ring $\mathbb{H}$.

1.6 Kreuzer's example

Following [Kreuzer 2 (1998)] we construct a semilinear 3-embedding of a certain vector space in a 3-dimensional vector space which induces a bijection between the Grassmannians of 2-dimensional subspaces. The dimension of the first vector space is greater than 3 and not necessarily finite.

Suppose that

$$E_1 \subset E_2 \subset E_3 \subset \cdots$$

is an infinite sequence of field extensions and

$$W_1, W_2, W_3, \ldots$$

is an infinite sequence of vector spaces over the same field F. We do not assume that E_i is a proper subfield of E_{i+1} and W_i is a proper subspace of W_{i+1}. Let

$$l_i : W_i \to E_i^3, \quad i = 1, 2, \ldots$$

be a sequence of semilinear 3-embeddings satisfying the following condition: the restriction of every l_{i+1} to W_i coincides with l_i. We define

$$W := \bigcup_{i \in \mathbb{N}} W_i, \quad E := \bigcup_{i \in \mathbb{N}} E_i$$

and denote by l the mapping of W to E^3 whose restriction to every W_i coincides with l_i. It is clear that l is a semilinear 3-embedding. The vector space W is finite-dimensional only in the case when there is a natural number i such that all W_j with $j \geq i$ are coincident. As in the previous subsection, we define the mapping

$$(l)_2 : \mathcal{G}_2(W) \to \mathcal{G}_2(E^3)$$

and establish that it is injective.

For every 2-dimensional subspace $U \subset E_i^3$ we denote by $\mathrm{d}_i(U)$ the dimension of the subspace of E_i^3 spanned by

$$U \cap l_i(W_i).$$

Then $\mathrm{d}_i(U)$ takes value from the set $\{0, 1, 2\}$. Let $\mathcal{A}_i$ and $\mathcal{B}_i$ be the sets formed by all 2-dimensional subspaces $U \subset E_i^3$ satisfying $\mathrm{d}_i(U) = 1$ and $\mathrm{d}_i(U) = 0$, respectively.

Now we construct the sequence of semilinear 3-embeddings described above. We take any field E_1 containing a proper subfield F. Let l_1 be the identity embedding of $W_1 := F^3$ in E_1^3. It follows from Corollary 1.2 that $\mathcal{A}_1 \cup \mathcal{B}_1$ is non-empty. Suppose that for a certain number i the embedding l_i is defined and the set $\mathcal{A}_i \cup \mathcal{B}_i$ is non-empty. Then l_{i+1} can be constructed in several steps.

(i). For every $U \in \mathcal{A}_i \cup \mathcal{B}_i$ we take two linearly independent vectors $x_U, y_U \in U$ such that

$$x_U \in U \cap l_i(W_i) \ \text{ if } \ U \in \mathcal{A}_i.$$

(ii). We define $E_{i+1} := E_i(T)$, where

$$T = \{t_U : U \in \mathcal{A}_i \cup \mathcal{B}_i\}$$

consists of algebraic or transcendental elements over E_i. We require that the degree of every $t_U \in T$ over $E_i(T \setminus \{t_U\})$ is not less than 3.

(iii). Consider a vector space W_{i+1} over F containing W_i and such that

$$\dim W_{i+1} = \dim W_i + |\mathcal{A}_i \cup \mathcal{B}_i|$$

(possibly $|\mathcal{A}_i \cup \mathcal{B}_i|$ is infinite). We take any base B_i of W_i and extend it to a base

$$B_{i+1} = B_i \cup \{w_U : U \in \mathcal{A}_i \cup \mathcal{B}_i\}$$

of W_{i+1}. Every vector $w \in W_{i+1}$ has the unique representation

$$w = w_0 + \sum_{U \in \mathcal{A}_i \cup \mathcal{B}_i} a_U w_U, \quad w_0 \in W_i, \ a_U \in F$$

and a_U is non-zero only for finitely many $U \in \mathcal{A}_i \cup \mathcal{B}_i$. We define

$$l_{i+1}(w) := l_i(w_0) + \sum_{U \in \mathcal{A}_i \cup \mathcal{B}_i} a_U(t_U x_U + t_U^2 y_U).$$

Lemma 1.4. *The mapping l_{i+1} is a semilinear 3-embedding of W_{i+1} in* E_{i+1}^3.

Proof. Let x, y, z be three distinct vectors of W_i. Suppose that $U \in \mathcal{A}_i \cup \mathcal{B}_i$, $c \in F \setminus \{0\}$ and the vectors

$$x, \; y, \; z + cw_U \tag{1.6}$$

are linearly independent in W_{i+1}. Then x and y are linearly independent. Since l_i is a semilinear 3-embedding of W_i in E_i^3, the vectors $l_i(x)$ and $l_i(y)$ are linearly independent in E_i^3. If the vectors

$$l_{i+1}(x) = l_i(x), \; l_{i+1}(y) = l_i(y), \; l_{i+1}(z + cw_U) = l_i(z) + c(t_U x_U + t_U^2 y_U)$$

are linearly dependent in E_{i+1}^3 then

$$\det[l_i(z) + c(t_U x_U + t_U^2 y_U), l_i(x), l_i(y)] =$$

$$\det[l_i(z), l_i(x), l_i(y)] + ct_U \det[x_U, l_i(x), l_i(y)] + ct_U^2 \det[y_U, l_i(x), l_i(y)] = 0.$$

The vectors $l_i(x), l_i(y), x_U, y_U$ belong to E_i^3 and

$$\det[l_i(z), l_i(x), l_i(y)], \;\; c\det[x_U, l_i(x), l_i(y)], \;\; c\det[y_U, l_i(x), l_i(y)]$$

are elements of E_i. By (ii), the degree of t_U over E_i is not less than 3 which implies that

$$\det[l_i(z), l_i(x), l_i(y)] = \det[x_U, l_i(x), l_i(y)] = \det[y_U, l_i(x), l_i(y)] = 0.$$

The vectors $l_i(x)$ and $l_i(y)$ are linearly independent and the latter equalities guarantee that each of the vectors $l_i(z), x_U, y_U$ is a linear combination of $l_i(x)$ and $l_i(y)$. Hence $U = \langle x_U, y_U \rangle$ is spanned by $l_i(x)$ and $l_i(y)$. Then $d_i(U) = 2$ which contradicts the assumption that $U \in \mathcal{A}_i \cup \mathcal{B}_i$. Thus l_{i+1} transfers (1.6) to a triple of linearly independent vectors of E_{i+1}^3.

Now we suppose that

$$x + aw_U, \; y + bw_U, \; z + cw_U \tag{1.7}$$

are linearly independent vectors in W_{i+1} and at least one of the scalars a, b is non-zero. Let N be the 3-dimensional subspace of W_{i+1} spanned by (1.7). Observe that N is spanned by $z + cw_U$ and some vectors $x', y' \in W_i$. It was established above that l_{i+1} transfers

$$x', \; y', \; z + cw_U$$

to a triple of linearly independent vectors. Then $l_{i+1}(N)$ spans a 3-dimensional subspace of E_{i+1}^3 which implies that l_{i+1} sends (1.7) to a triple of linearly independent vectors of E_{i+1}^3. Therefore, the restriction of l_{i+1} to the subspace of W_{i+1} spanned by W_i and w_U is a semilinear 3-embedding.

Let $\mathcal{X}$ be a proper subset of $\mathcal{A}_i \cup \mathcal{B}_i$. Denote by W' the subspace of W_{i+1} spanned by W_i and all vectors w_Q such that $Q \in \mathcal{X}$. Suppose that the restriction of l_{i+1} to W' is a semilinear 3-embedding and consider any $U \in \mathcal{A}_i \cup \mathcal{B}_i$ which does not belong to $\mathcal{X}$. Using the above arguments, we show that the restriction of l_{i+1} to the subspace of W_{i+1} spanned by W' and w_U is a semilinear 3-embedding (we exploit the fact that the degree of t_U over $E_i(T \setminus \{t_U\})$ is not less than 3). After that we can apply Zorn's lemma and establish that l_{i+1} is a semilinear 3-embedding. □

So, the semilinear 3-embedding l_{i+1} is constructed. Note that W_i is a proper subspace of W_{i+1}. If the set $\mathcal{A}_{i+1} \cup \mathcal{B}_{i+1}$ is non-empty then we apply the above arguments to l_{i+1} and construct l_{i+2}.

In the case when $\mathcal{A}_i \cup \mathcal{B}_i$ is empty for a certain $i \geq 2$, we have $\mathrm{d}_i(U) = 2$ for every 2-dimensional subspace $U \subset E_i^3$. We choose two linearly independent vectors x', y' belonging to $U \cap l_i(W_i)$. Then $l_i^{-1}(x')$ and $l_i^{-1}(y')$ are linearly independent vectors of W_i and l_i transfers the 2-dimensional subspace spanned by them to a subset of E_i^3 spanning U. This means that the mapping

$$(l_i)_2 : \mathcal{G}_2(W_i) \to \mathcal{G}_2(E_i^3)$$

is bijective. We have $\dim W_i \geq \dim W_2 > 3$ (recall that $\mathcal{A}_1 \cup \mathcal{B}_1$ is non-empty) and the semilinear 3-embedding l_i is as required.

From this moment we will suppose that $\mathcal{A}_i \cup \mathcal{B}_i$ is non-empty for every i. In this case, we get an infinite sequence of semilinear 3-embeddings $l_i : W_i \to E_i^3$. Since every W_i is a proper subspace of W_{i+1}, the vector space W is infinite-dimensional. Now we show that the semilinear 3-embedding $l : W \to E^3$ is as required, i.e. the mapping $(l)_2$ is bijective.

We can consider every 2-dimensional subspace $U \subset E_i^3$ as a subset of E_{i+1}^3 and write U' for the subspace of E_{i+1}^3 spanned by U.

Lemma 1.5. *For every integer i and every 2-dimensional subspace $U \subset E_i^3$ we have*

$$\mathrm{d}_{i+1}(U') \geq 1 \ \ if \ \ \mathrm{d}_i(U) < 2$$

and

$$\mathrm{d}_{i+1}(U') = 2 \ \ if \ \ \mathrm{d}_i(U) = 1.$$

Proof. If $\mathrm{d}_i(U) < 2$ then $U \in \mathcal{A}_i \cup \mathcal{B}_i$ and we have

$$l_{i+1}(w_U) = t_U x_U + t_U^2 y_U \in U'$$

which means that $d_{i+1}(U') \geq 1$. If $d_i(U) = 1$, i.e. $U \in \mathcal{A}_i$, then

$$x_U \in U \cap l_i(W_i) \subset U' \cap l_{i+1}(W_{i+1}).$$

This implies that $d_{i+1}(U') = 2$, since the vectors x_U and $t_U x_U + t_U^2 y_U$ are linearly independent. $\qquad\qquad\square$

Lemma 1.6. *Every 2-dimensional subspace $S \subset E^3$ is spanned by $S \cap l(W)$.*

Proof. We choose linearly independent vectors $x, y \in S$. There is an integer i such that these vectors both belong to E_i^3. Then

$$U := S \cap E_i^3$$

is a 2-dimensional subspace of E_i^3. In the case when $d_i(U) = 2$, the statement is obvious. If $d_i(U) = 1$ then, by Lemma 1.5, we have $d_{i+1}(U') = 2$ which implies that S is spanned by $S \cap l(W)$. Similarly, if $d_i(U) = 0$ then $d_{i+1}(U') \geq 1$. The latter guarantees that $d_{i+2}(U'') = 2$ and we get the claim. $\qquad\qquad\square$

It follows from Lemma 1.6 that $(l)_2$ is bijective.

1.7   Duality

Let V be a left vector space over a division ring R. Recall that linear mappings of V to R are called *linear functionals*. For linear functionals $\alpha : V \to R$ and $\beta : V \to R$ the sum $(\alpha + \beta) : V \to R$ is defined as follows

$$(\alpha + \beta)(x) := \alpha(x) + \beta(x) \quad \forall\, x \in V.$$

For a scalar $a \in R$ and a linear functional $\alpha : V \to R$ we define the product $(\alpha a) : V \to R$ by the formula

$$(\alpha a)(x) := \alpha(x)a \quad \forall\, x \in V.$$

The set of all linear functionals together with the above defined operations is a right vector space over R. The associated left vector space over the opposite division ring R^* is denoted by V^* and called the *dual vector space*.

In what follows for all $x \in V$ and $x^* \in V^*$ we will write $x^* \cdot x$ instead of $x^*(x)$.

Suppose that $\dim V = n$ is finite. Let $B = \{x_1, \dots, x_n\}$ be a base of V. Denote by B^* the set formed by the linear functionals $x_1^*, \dots, x_n^*$ satisfying

$$x_i^* \cdot x_j = \delta_{ij},$$

where δ_{ij} is Kronecker symbol. This is a base of V^*. Hence $\dim V^* = n$. The base B^* is called *dual* to the base B.

Consider the second dual vector space V^{**}. For every vector $x \in V$ the mapping α_x transferring every $x^* \in V^*$ to $x^* \cdot x$ is a linear functional. The correspondence $x \to \alpha_x$ is a linear isomorphism between V and V^{**}. In what follows we will identify V^{**} with V.

If X is a subset of V then

$$X^0 := \{x^* \in V^* : x^* \cdot x = 0 \quad \forall\, x \in X\}$$

is a subspace of V^*. This subspace is known as the *annihilator* of X. Since X^0 is the kernel of the linear mapping of V^* to $\langle X \rangle^*$ which sends every linear functional to its restriction to $\langle X \rangle$, the dimension of X^0 is equal to $n - \dim \langle X \rangle$. Similarly, for every subset $Y \subset V^*$ the subspace

$$Y^0 := \{x \in V : x^* \cdot x = 0 \quad \forall\, x^* \in Y\}$$

is called the *annihilator* of Y. Its dimension is equal to $n - \dim \langle Y \rangle$. The following assertions are fulfilled:

- $X^{00} = \langle X \rangle$ for every subset $X \subset V$ and if S is a subspace of V then $S^{00} = S$.
- For every subspace $S \subset V$ the annihilator S^0 can be naturally identified with $(V/S)^*$.

Lemma 1.7. *If B is a base of V and S is a subspace spanned by a subset of B then S^0 is spanned by a subset of the dual base B^*.*

Proof. Suppose that $B = \{x_1, \ldots, x_n\}$. If S is spanned by $x_1, \ldots, x_k$ then S^0 is spanned by $x_{k+1}^*, \ldots, x_n^*$, where x_i^* is the vector of B^* satisfying $x_i^* \cdot r_j - \delta_{ij}$. $\square$

Lemma 1.8. *For any collection of subspaces $S_1, \ldots, S_k \subset V$ we have*

$$(S_1 + \cdots + S_k)^0 = (S_1)^0 \cap \cdots \cap (S_k)^0,$$

$$(S_1 \cap \cdots \cap S_k)^0 = (S_1)^0 + \cdots + (S_k)^0.$$

Proof. Easy verification. $\square$

The annihilator mapping of $\mathcal{G}(V)$ to $\mathcal{G}(V^*)$ is a bijection transferring every $\mathcal{G}_k(V)$ to $\mathcal{G}_{n-k}(V^*)$. This bijection is inclusion reversing, i.e.

$$S \subset U \iff U^0 \subset S^0$$

for any subspaces $S, U \subset V$. In particular, it sends $[S, U]_k$ to $[U^0, S^0]_{n-k}$. Since $S^{00} = S$ for every subspace $S \subset V$, the inverse mapping is the annihilator mapping of $\mathcal{G}(V^*)$ to $\mathcal{G}(V)$.

Throughout the book we will use the following notation: for every subset $\mathcal{X} \subset \mathcal{G}(V)$ we denote by $\mathcal{X}^0$ the subset of $\mathcal{G}(V^*)$ formed by the annihilators of the elements from $\mathcal{X}$.

Let V' be a left vector space over a division ring R'. Suppose that $u : V \to V'$ is a semilinear isomorphism and σ is the associated isomorphism of R to R'. Then $\dim V' = n$. For every $x^* \in V'^*$ the mapping

$$x \to \sigma^{-1}(x^* \cdot u(x))$$

is a linear functional of V. This functional is denoted by $u^*(x^*)$. An easy verification shows that the *adjoint* mapping

$$u^* : V'^* \to V^*$$

is a (σ^{-1})-linear isomorphism. The inverse mapping

$$(u^*)^{-1} : V^* \to V'^*$$

is a σ-linear isomorphism of V^* to V'^*. It is called the *contragradient* of u and denoted by $\breve{u}$. The equalities

$$(u^*)^{-1} = (u^{-1})^* \quad \text{and} \quad u^{**} = u$$

show that $\breve{\breve{u}} = u$.

Example 1.20. Let $u : V \to V^*$ be a semilinear isomorphism and let σ be the associated isomorphism of R to R^*. Then u^* is a (σ^{-1})-linear isomorphism of V to V^*. If u is self-adjoint, i.e. $u = u^*$, then σ^2 is identity. Conversely, for any isomorphism $\sigma : R \to R^*$ such that σ^2 is identity there exist self-adjoint σ-linear isomorphisms of V to V^*. For example, if $x_1, \dots, x_n$ is a base of V, $x_1^*, \dots, x_n^*$ is the dual base of V^* and $x_i^* \cdot x_j = \delta_{ij}$ then the σ-linear isomorphism of V to V^* sending every x_i to x_i^* is self-adjoint. In particular, if R is a field then self-adjoint linear isomorphisms of V to V^* exist; if R is the quaternion division ring $\mathbb{H}$ then there exist self-adjoint semilinear isomorphisms of V to V^* associated to the conjugate mapping.

Proposition 1.10. *The contragradient mapping is an isomorphism of* $\Gamma L(V)$ *to* $\Gamma L(V^*)$ *transferring* $GL(V)$ *to* $GL(V^*)$.

Proof. If $u, v \in \Gamma L(V)$ then $(uv)^* = v^* u^*$ which implies that the contragradient of uv coincides with $\breve{u}\breve{v}$. The bijectivity follows from the equality $\breve{\breve{u}} = u$. $\qquad\square$

Proposition 1.11. *If $u : V \to V'$ is a semilinear isomorphism then*

$$\breve{u}(U) = u(U^0)^0$$

for every subspace $U \subset V^*$, in other words, for every $k \in \{1, \ldots, n-1\}$ the mapping of $\mathcal{G}_k(V^*)$ to $\mathcal{G}_k(V'^*)$ sending every U to $u(U^0)^0$ coincides with $(\breve{u})_k$.

Proof. By the definition of the adjoint mapping, for $x \in V$ and $x^* \in V'^*$ we have

$$u^*(x^*) \cdot x = \sigma^{-1}(x^* \cdot u(x)) \quad \text{and} \quad \sigma(u^*(x^*) \cdot x) = x^* \cdot u(x).$$

If $x^* = \breve{u}(y^*)$ for a certain $y^* \in V^*$ then

$$\sigma(y^* \cdot x) = \breve{u}(y^*) \cdot u(x)$$

and

$$y^* \cdot x = 0 \iff \breve{u}(y^*) \cdot u(x) = 0.$$

The latter implies the required equality. $\qquad\square$

Suppose that $\dim V' = n$ and $u : V \to V'$ is a strong σ-linear embedding. Then $u(V)$ is a left n-dimensional vector space over $\sigma(R)$ and u is a σ-linear isomorphism of V to this vector space. Every linear functional of $u(V)$ can be uniquely extended to a linear functional of V'. This extension is a strong semilinear embedding of $u(V)^*$ in V'^* over the identity homomorphism of $\sigma(R)^*$ to R'^*. Thus the contragradient of the semilinear isomorphism $u : V \to u(V)$ can be considered as a strong σ-linear embedding of V^* in V'^*. This embedding will be called the *contragradient* of the embedding $u : V \to V'$ and denoted by $\breve{u}$.

Proposition 1.11 can be generalized as follows.

Proposition 1.12. *If $\dim V = \dim V' = n$ and $u : V \to V'$ is a strong semilinear embedding then*

$$\langle \breve{u}(U) \rangle = u(U^0)^0$$

for every subspace $U \subset V^*$, in other words, for every $k \in \{1, \ldots, n-1\}$ the mapping of $\mathcal{G}_k(V^*)$ to $\mathcal{G}_k(V'^*)$ sending every U to $u(U^0)^0$ coincides with $(\breve{u})_k$.

Proof. Let $U \in \mathcal{G}_k(V^*)$. By Proposition 1.11, the annihilator of $u(U^0)$ in $u(V)^*$ coincides with $\breve{u}(U)$. Thus $\breve{u}(U)$ is a k-dimensional subspace of $u(V)^*$ contained in $u(U^0)^0$ which is a k-dimensional subspace of V'^*. Then $\langle \breve{u}(U) \rangle$ coincides with $u(U^0)^0$. $\qquad\square$

Remark 1.7. Since $u(U^0)^0 = \langle u(U^0)\rangle^0$, we have

$$(\breve{u})_k = \mathrm{A}'(u)_{n-k}\mathrm{A},$$

where A is the annihilator mapping of $\mathcal{G}_k(V^*)$ to $\mathcal{G}_{n-k}(V)$ and A′ is the annihilator mapping of $\mathcal{G}_{n-k}(V')$ to $\mathcal{G}_k(V^*)$.

1.8 Characterization of strong semilinear embeddings

Let V and V' be left vector spaces over division rings R and R', respectively. Suppose that $\dim V = n$ is finite. Every strong semilinear embedding $l : V \to V'$ has the following property:

(GL) for every $u \in \mathrm{GL}(V)$ there is $u' \in \mathrm{GL}(V')$ such that $lu = u'l$.

Indeed, if $u \in \mathrm{GL}(V)$ and $x_1, \ldots, x_n$ is a base of V then

$$\{l(x_1), \ldots, l(x_n)\} \text{ and } \{lu(x_1), \ldots, lu(x_n)\}$$

are independent subsets of V'. Any $u' \in \mathrm{GL}(V')$ transferring every $l(x_i)$ to $lu(x_i)$ is as required.

Remark 1.8. It is not difficult to prove that for any strong σ-linear embedding $l : V \to V'$ the following two conditions are equivalent:

- for every $u \in \Gamma\mathrm{L}(V)$ there is $u' \in \Gamma\mathrm{L}(V)$ such that $lu = u'l$,
- for every $\gamma \in \mathrm{Aut}(R)$ there is $\gamma' \in \mathrm{Aut}(R')$ such that $\sigma\gamma = \gamma'\sigma$.

The second condition does not hold, for example, if σ is a homomorphism of $\mathbb{C}$ to $\mathbb{H}$ leaving fixed each real number (the restriction of every automorphism of $\mathbb{H}$ to $\mathbb{R}$ is identity, but there are automorphisms of $\mathbb{C}$ whose restrictions to $\mathbb{R}$ are non-identity). Another one example is the identity homomorphisms of the fields from Examples 1.2 and 1.3 to $\mathbb{R}$.

An arbitrary (not necessarily semilinear) mapping $l : V \to V'$ satisfying the condition (GL) will be called a GL-*mapping*.

Example 1.21. Let $l : V \to V'$ be a mapping whose restriction to $V \setminus \{0\}$ is constant, i.e. there is $x' \in V'$ such that $l(x) = x'$ for every non-zero vector $x \in V$. The equality $lu = u'l$ holds for every $u \in \mathrm{GL}(V)$ and any $u' \in \mathrm{GL}(V')$ which leaves fixed $l(0)$ and x' (we do not require that $l(0) = 0$).

A GL-mapping of V to V' is said to be *non-trivial* if its restriction to $V \setminus \{0\}$ is non-constant. We present a few examples of non-trivial GL-mappings which are not strong semilinear embeddings.

Example 1.22. Consider R as a 1-dimensional left vector space over itself. Every linear automorphism of this vector space is of type $x \to xa$ with $a \in R \setminus \{0\}$. Thus $l : R \to R$ is a GL-mapping if for every $a \in R \setminus \{0\}$ there exists $a' \in R \setminus \{0\}$ such that $l(xa) = l(x)a'$ for all $x \in R$. The latter holds for any mapping $l : R \to R$ such that $l(0) = 0$ and $l|_{R \setminus \{0\}}$ is an endomorphism of the multiplicative group of R. This mapping is not necessarily additive. For example, R is commutative and $l|_{R \setminus \{0\}}$ sends every x to x^{-1}.

We will use the following notation: for every mapping $l : V \to V'$ we denote by V_l the subspace of V' spanned by $l(V)$.

Example 1.23. Let $l : V \to V'$ be a strong semilinear embedding. Suppose that $n < \dim V'$ and take any vector $x' \in V'$ which does not belong to the n-dimensional subspace V_l. Consider the mapping $s : V \to V'$ defined as follows

$$s(x) = \begin{cases} l(x) & x \neq 0 \\ x' & x = 0. \end{cases}$$

This is a GL-mapping. Indeed, for every $u \in \mathrm{GL}(V)$ there exists $\bar{u} \in \mathrm{GL}(V_l)$ satisfying $lu = \bar{u}l$ and we have $su = u's$ for every $u' \in \mathrm{GL}(V')$ which is an extension of $\bar{u}$ and leaves fixed x'.

Example 1.24 (H. Havlicek). Suppose that R is a finite field and $\dim V' = |V|$. Consider any bijection l of V to a base B' of V'. If u is a bijective transformation of V then lul^{-1} is a permutation on B'. Since B' is a base of V', we can extend lul^{-1} to a certain $u' \in \mathrm{GL}(V')$. Then $lu = u'l$. Thus l is a GL-mapping.

Example 1.25. Suppose that $l : V \to V'$ is a σ-linear bijection. Note that l is not necessarily a semilinear isomorphism, see Example 1.14. The inverse mapping l^{-1} is additive and we have

$$l^{-1}(\sigma(a)x') = al^{-1}(x')$$

for every $x' \in V'$ and $a \in R$. Using these properties, we establish that for every $u' \in \mathrm{GL}(V')$ the mapping $l^{-1}u'l$ is a linear automorphism of V. This means that l^{-1} is a GL-mapping.

There is the following characterization of strong semilinear embeddings.

Theorem 1.2 ([Pankov (2013)]). *Suppose that $n \geq 3$ and $l : V \to V'$ is a non-trivial GL-mapping. If*

$$\dim V_l \leq n \qquad\qquad (1.8)$$

then l is a strong semilinear embedding.

For $n = 2$ the same statement is not proved and, by Example 1.22, it fails for $n = 1$.

Examples 1.23–1.25 show that the condition (1.8) in Theorem 1.2 cannot be omitted.

If $l : V \to V'$ is a non-trivial GL-mapping then for every $u \in \mathrm{GL}(V)$ there is unique $\overline{u} \in \mathrm{GL}(V_l)$ such that $lu = \overline{u}l$. It is not difficult to prove that the mapping $u \to \overline{u}$ is a homomorphism of $\mathrm{GL}(V)$ to $\mathrm{GL}(V_l)$. We denote this homomorphism by H_l.

All automorphisms and isomorphisms of classical groups are known [Dieudonné (1971); O'Meara (1974)]. For example, we have the following.

Theorem 1.3. *There are precisely the following two types of automorphisms of the group $\mathrm{GL}(V)$:*

- *$u \to \alpha(u)lul^{-1}$, where $l \in \Gamma\mathrm{L}(V)$,*
- *$u \to \alpha(u)s^{-1}\breve{u}s$, where $s : V \to V^*$ is a semilinear isomorphism and $\breve{u}$ is the contragradient of u;*

in each of these cases, α is a homomorphism of $\mathrm{GL}(V)$ to the center of R.

Proof. See [Dieudonné (1971), Section IV.1]. □

Homomorphisms between linear groups are not described in the general case, but there is the following.

Theorem 1.4 ([Dicks and Hartley (1991); Zha (1996)]). *Suppose that the division rings R and R' are of the same characteristic, R' is a finite-dimensional vector space over the center and $\dim V' = n \geq 2$. Then for every non-trivial homomorphism $H : \mathrm{SL}(V) \to \mathrm{GL}(V')$ one of the following possibilities is realized:*

- *there is a strong semilinear embedding $l : V \to V'$ such that H is the restriction of H_l to $\mathrm{SL}(V)$,*
- *there is a strong semilinear embedding $s : V \to V'^*$ such that H is the composition of the restriction of H_s to $\mathrm{SL}(V)$ and the contragradient isomorphism of $\mathrm{GL}(V'^*)$ to $\mathrm{GL}(V')$.*

Remark 1.9. Recall that $\mathrm{SL}(V)$ is the subgroup of $\mathrm{GL}(V)$ generated by all transvections, or equivalently, $\mathrm{SL}(V)$ consists of all linear automorphisms whose determinant is equal to 1 [Artin (1957), Section IV.1] and [Dieudonné (1971), Section II.1].

Theorem 1.2 easy follows from Theorem 1.4 in the case when $\dim V_l = n$, the division rings R and R' are of the same characteristic and R' is a finite-dimensional vector space over the center. By Proposition 1.3, there are division rings which are infinity-dimensional vector spaces over the centers.

Theorem 1.2 will be proved in the next chapter. In Chapter 3 we will need the following weak version of this result.

Proposition 1.13. *Every non-zero semilinear GL-mapping of V to V' is a strong semilinear embedding.*

In the case when $n \geq 3$, this is a direct consequence of Theorem 1.2. We give a simple proof of Proposition 1.13 non-related to Theorem 1.2.

Proof of Proposition 1.13. The statement is trivial for $n = 1$ and we suppose that $n \geq 2$.

Let $l : V \to V'$ be a non-zero semilinear GL-mapping. Then for every $u \in \mathrm{GL}(V)$ there exists $u' \in \mathrm{GL}(V')$ satisfying $lu = u'l$. If $l(x) = 0$ for a certain non-zero vector $x \in V$ then for every non-zero vector $y \in V$ there exists $u \in \mathrm{GL}(V)$ transferring x to y and we have

$$l(y) = lu(x) = u'l(x) = 0$$

which contradicts the assumption that l is non-zero. So, l is injective and it is sufficient to show that l sends every base of V to an independent subset of V'.

Suppose that $x_1, \ldots, x_n$ is a base of V such that

$$l(x_n) = \sum_{i=1}^{n-1} a_i l(x_i).$$

For every $u \in \mathrm{GL}(V)$ we have

$$lu(x_n) = u'l(x_n) = \sum_{i=1}^{n-1} a_i u'l(x_i) = \sum_{i=1}^{n-1} a_i lu(x_i).$$

Consider $v \in \mathrm{GL}(V)$ such that

$$u(x_i) = v(x_i) \text{ if } i \leq n - 1 \text{ and } u(x_n) \neq v(x_n).$$

Then

$$lv(x_n) = \sum_{i=1}^{n-1} a_i lv(x_i) = \sum_{i=1}^{n-1} a_i lu(x_i) = lu(x_n)$$

which is impossible, since l is injective.   □

Chapter 2

Projective Geometry and linear codes

First we discuss the geometrical characterization of semilinear mappings known as the Fundamental Theorem of Projective Geometry. The classical version of this theorem states that every collineation between projective spaces is induced by a semilinear isomorphism of the associated vector spaces [Artin (1957); Baer (1952); Dieudonné (1971)]. We present Faure–Frölicher–Havlicek's result [Faure and Frölicher (1994); Havlicek (1994)] characterizing a more general class of semilinear mappings. We use this statement to prove Theorem 1.2. Also, it will be exploited in Chapter 3 when we describe isometric embeddings of Grassmann graphs.

One of the main objects of classical Projective Geometry is a simplex. An m-simplex consists of $m + 1$ distinct points, it is not an independent subset, but its m-element subsets are independent. We generalize this notion in the following way: a subset in a projective space will be called m-independent if it contains at least m points and its m-element subsets are independent.

In this chapter we give only several examples and establish some elementary properties of m-independent subsets. There are interesting examples of these subsets — so-called PGL-subsets; every permutation on such a subset can be extended to an element of the projective linear group. We have a simple classification of PGL-subsets in the projective spaces over fields and for the non-commutative case the same is an open problem. In Chapter 4 we will use m-independent subsets in the description of isometric embeddings of Johnson graphs in Grassmann graphs.

Linear codes, i.e. subspaces of vector spaces over finite fields, will be considered in the second part of the chapter. Following [Tsfasman, Vlăduţ and Nogin (2007)] we identify them with projective systems, i.e. collections of not necessarily distinct points in the projective spaces over finite fields.

So, it will be natural to interpret m-independent subsets and PGL-subsets in terms of linear codes.

Also, this geometrical approach provides a simple proof of classical MacWilliams theorem [MacWilliams (1961)] which states that the class of semilinear monomial isomorphisms, i.e. isomorphisms of linear codes, coincides with the class of weight preserving semilinear isomorphisms. The same holds for all vector spaces over division rings and we present this generalized MacWilliams theorem before linear codes.

We consider linear codes for the following two reasons. The classification of isometric embeddings of Johnson graphs in Grassmann graphs, if the associated vector space is over a finite field, is equivalent to the classification of linear codes of special type (Chapter 4). Chow's theorem (concerning automorphisms of Grassmann graphs) is related to the description of the automorphism groups of Grassmann codes (Chapter 6).

2.1   Projective spaces

Let $\mathcal{P}$ be a non-empty set whose elements are said to be *points* and let $\mathcal{L}$ be a family of proper subsets of $\mathcal{P}$ called *lines*. The pair $\Pi = (\mathcal{P}, \mathcal{L})$ is a *projective space* if the following axioms hold:

(1) each line contains at least three points;
(2) for any two distinct points $p, q \in \mathcal{P}$ there is precisely one line containing them, this line will be denoted by pq;
(3) if a, b, c, d are distinct points such that the lines ab and cd have a non-empty intersection then the lines ac and bd have a non-empty intersection.

The axiom (2) guarantees that the intersection of two distinct lines contains at most one point. We say that three or more points are *collinear* if there is a line containing them; otherwise, these points are said to be *non-collinear*.

A subset $\mathcal{S} \subset \mathcal{P}$ is a *subspace* of Π if for any two distinct points $p, q \in \mathcal{S}$ the line pq is contained in $\mathcal{S}$. The empty set, one-point sets, lines and $\mathcal{P}$ are subspaces. The intersection of any collection of subspaces is a subspace.

Let $\mathcal{X}$ be a subset of $\mathcal{P}$. The minimal subspace containing $\mathcal{X}$, i.e. the intersection of all subspaces containing $\mathcal{X}$, is called *spanned* by $\mathcal{X}$ and denoted by $\langle \mathcal{X} \rangle$. We say that $\mathcal{X}$ is *independent* if the subspace $\langle \mathcal{X} \rangle$ cannot be spanned by a proper subset of $\mathcal{X}$. For example, the subset formed by three or more collinear points is not independent.

Let $\mathcal{S}$ be a subspace of Π (possibly $\mathcal{S}$ coincides with $\mathcal{P}$). An independent subset $\mathcal{X} \subset \mathcal{S}$ is a *base* of $\mathcal{S}$ if $\langle \mathcal{X} \rangle$ coincides with $\mathcal{S}$. We define the *dimension* of $\mathcal{S}$ as the smallest cardinality α such that $\mathcal{S}$ is spanned by a subset of cardinality $\alpha + 1$. Points and lines are 0-dimensional and 1-dimensional subspaces, respectively. The dimension of the empty set is equal to -1. Every subspace of dimension not less than 2 (together with the lines contained in it) is a projective space. A 2-dimensional projective space is called a *projective plane*. Projective planes are characterized by the following property: any two distinct lines have a non-empty intersection.

We say that projective spaces $\Pi = (\mathcal{P}, \mathcal{L})$ and $\Pi' = (\mathcal{P}', \mathcal{L}')$ are *isomorphic* if there is a bijection $f : \mathcal{P} \to \mathcal{P}'$ transferring any triple of collinear points to a triple of collinear points and any triple of non-collinear points to a triple of non-collinear points, or equivalently, $f(\mathcal{L}) = \mathcal{L}'$. Bijections satisfying this condition are known as *collineations*. It is clear that every collineation transfers subspaces to subspaces and isomorphic projective spaces are of the same dimension.

A bijection of $\mathcal{P}$ to $\mathcal{P}'$ is a *semicollineation* if it transfers any triple of collinear points to a triple of collinear points; in other words, semicollineations are bijections which map lines to subsets of lines.

Example 2.1. Let V be a left vector space over a division ring. Suppose that $\dim V = n \geq 3$. Denote by Π_V the projective space whose point set is $\mathcal{G}_1(V)$ and whose lines are all subsets of type

$$\mathcal{G}_1(S), \quad S \in \mathcal{G}_2(V).$$

Then every subspace of Π_V is a subset of type $\mathcal{G}_1(U)$, where U is a subspace of V, and the dimension of this subspace is equal to $\dim U - 1$. The projective space Π_V is $(n - 1)$-dimensional.

The following fact is well-known.

Theorem 2.1. *Every projective space of dimension not less than 3 is isomorphic to the projective space associated to a vector space over a division ring.*

For projective planes the same fails. There exist non-desarguesian projective planes.

Example 2.2. As in the previous example, we suppose that V is a left vector space over a division ring and $\dim V = n \geq 3$. Consider the *dual* projective space Π_V^* whose point set is $\mathcal{G}_{n-1}(V)$ and whose lines are all subsets of type

$$[U\rangle_{n-1}, \quad U \in \mathcal{G}_{n-2}(V).$$

The annihilator mapping of $\mathcal{G}_{n-1}(V)$ to $\mathcal{G}_1(V^*)$ is a collineation of Π_V^* to Π_{V^*}. This means that every subspace of Π_V^* is a subset of type $[U\rangle_{n-1}$, where U is a subspace of V. The dimension of this subspace is equal to $n - \dim U - 1$ and the projective space Π_V^* is $(n-1)$-dimensional.

If V is a left 2-dimensional vector space over a division ring then the Grassmannian $\mathcal{G}_1(V)$ is known as the *projective line* associated to V. We will suppose that this line is spanned by any pair of distinct points and every subset containing more than two points is not independent. So, our line is similar to lines of projective spaces.

2.2   Fundamental Theorem of Projective Geometry

Let V and V' be left vector spaces over division rings. Suppose that $\dim V = n$ and $\dim V' = n'$ both are finite and not less than 3.

Let $l : V \to V'$ be a semilinear injection. Consider the associated mapping

$$(l)_1 : \mathcal{G}_1(V) \to \mathcal{G}_1(V'),$$

see Section 1.5. For any distinct $P, P_1, P_2 \in \mathcal{G}_1(V)$ satisfying $P \subset P_1 + P_2$, i.e. P, P_1, P_2 is a triple of collinear points in Π_V, we have

$$\langle l(P) \rangle \subset \langle l(P_1) \rangle + \langle l(P_2) \rangle.$$

If $\langle l(P_1) \rangle = \langle l(P_2) \rangle$ then $(l)_1$ transfers the line $\mathcal{G}_1(P_1 + P_2)$ to a point. If $\langle l(P_1) \rangle$ and $\langle l(P_2) \rangle$ are distinct then the restriction of $(l)_1$ to this line is an injection to the line $\mathcal{G}_1(\langle l(P_1 + P_2) \rangle)$.

The following version of the Fundamental Theorem of Projective Geometry was obtained in [Faure and Frölicher (1994)] and [Havlicek (1994)], independently.

Theorem 2.2. *Let $f : \mathcal{G}_1(V) \to \mathcal{G}_1(V')$ be a mapping satisfying the following condition: if $P, P_1, P_2 \in \mathcal{G}_1(V)$ then*

$$P \subset P_1 + P_2 \implies f(P) \subset f(P_1) + f(P_2).$$

If the image $f(\mathcal{G}_1(V))$ is not contained in a line of $\Pi_{V'}$ then f is induced by a semilinear injection of V to V'.

The proof of this statement is a modification of the proof of the classical Fundamental Theorem of Projective Geometry [Artin (1957); Baer (1952)]. For this reason, we only sketch it and refer [Faure (2002)] or [Faure and Frölicher (2000), Theorem 10.1.3] or [Pankov (2010), Theorem 1.4] for the details.

Sketch of proof. Since $f(\mathcal{G}_1(V))$ is not contained in a line of $\Pi_{V'}$, there exist non-zero vectors $x_1, x_2, x_3 \in V$ such that

$$f(\langle x_1 \rangle), \ f(\langle x_2 \rangle), \ f(\langle x_3 \rangle)$$

are non-collinear points in $\Pi_{V'}$. The first step is to establish the existence of non-zero vectors $y_1, y_2, y_3 \in V'$ satisfying

$$f(\langle x_i \rangle) = \langle y_i \rangle \ \text{ and } \ f(\langle x_i + x_j \rangle) = \langle y_i + y_j \rangle.$$

For every non-zero vector $x \in V$ there are at least two numbers $i \in \{1, 2, 3\}$ such that $f(\langle x \rangle) \neq \langle y_i \rangle$. The point $\langle x_i + x \rangle$ is on the line joining $\langle x_i \rangle$ and $\langle x \rangle$. Then

$$f(\langle x_i + x \rangle) \subset \langle y_i \rangle + f(\langle x \rangle).$$

Since the restriction of f to a line is injective or constant and $f(\langle x \rangle) \neq \langle y_i \rangle$,

$$f(\langle x \rangle), f(\langle x_i + x \rangle), \langle y_i \rangle$$

are distinct points on a common line. This implies the existence of a non-zero vector $l(x) \in f(\langle x \rangle)$ such that

$$f(\langle x_i + x \rangle) = \langle y_i + l(x) \rangle.$$

We need to show that this vector does not depend on the choice of a number $i \in \{1, 2, 3\}$ satisfying $f(\langle x \rangle) \neq \langle y_i \rangle$.

Next, we set $l(0) := 0$ and establish that the mapping $l : V \to V'$ is semilinear. Since every $l(x)$ is a non-zero vector belonging to $f(\langle x \rangle)$, this mapping is a semilinear injection and $f = (l)_1$. □

If $l : V \to V'$ is a semilinear 2-embedding then $(l)_1$ is injective. We will use the following consequence of Theorem 2.2.

Corollary 2.1. *Let $f : \mathcal{G}_1(V) \to \mathcal{G}_1(V')$ be an injection transferring every line of Π_V to a subset in a line of $\Pi_{V'}$. If the image $f(\mathcal{G}_1(V))$ is not contained in a line of $\Pi_{V'}$ then f is induced by a semilinear 2-embedding of V in V'.*

Note that there is a mapping between projective spaces which transfers lines to subsets of lines and whose image is not contained in a line, but this mapping is not induced by a semilinear mapping. Every such mapping is non-injective.

Example 2.3. Consider the projective plane associated to a 3-dimensional vector space. In this plane we take any line l and a point p which does not belong to this line. Let f be the transformation of the plane defined by the following conditions:

- f leaves fixed p and all points on the line l,
- if a point q does not belong to l and is different from p then $f(q)$ is the intersecting point of the lines l and pq.

The image of our mapping is $l \cup \{p\}$. We check that f sends lines to subsets of lines. Let s be a line. If $p \notin s$ then $f(s)$ coincides with l. If $p \in s$ then $f(s)$ is a 2-element subset of s, i.e. the restriction of f to s is not injective or constant. The latter means that f cannot be induced by a semilinear transformation of the associated vector space.

If f is a collineation of Π_V to $\Pi_{V'}$ then Theorem 2.2 implies the existence of a semilinear injection $l : V \to V'$ such that $f = (l)_1$. Since f transfers any triple of non-collinear points to a triple of non-collinear points, l is a semilinear 3-embedding. We apply Proposition 1.8 and get the classical version of the Fundamental Theorem of Projective Geometry.

Corollary 2.2. *Every collineation of Π_V to $\Pi_{V'}$ is induced by a semilinear isomorphism of V to V'.*

Also, Corollary 2.1 together with Proposition 1.8 give the following.

Corollary 2.3. *If $n \le n'$ then every semicollineation of Π_V to $\Pi_{V'}$ is a collineation.*

There is a semicollineation of a 4-dimensional projective space to a non-desarguesian projective plane [Ceccherini (1967)]. Since every semi-collineation of Π_V to $\Pi_{V'}$ is induced by a semilinear 2-embedding of V in V', Problem 1.1 is closely related to the following classical problem.

Problem 2.1. Is there a semicollineation of Π_V to $\Pi_{V'}$ if $n > n'$? In other words, is there a semicollineation which is not a collineation?

If $l : V \to V'$ is a semilinear isomorphism then $n = n'$ and $(l)_{n-1}$ is a collineation between the dual projective spaces Π_V^* and $\Pi_{V'}^*$. Using Proposition 1.11 we prove the dual version of the Fundamental Theorem of Projective Geometry.

Corollary 2.4. *Every collineation of Π_V^* to $\Pi_{V'}^*$ is induced by a semilinear isomorphism of V to V'.*

Proof. Let f be a collineation of Π_V^* to $\Pi_{V'}^*$. The mapping of $\mathcal{G}_1(V^*)$ to $\mathcal{G}_1(V'^*)$ transferring every P to $f(P^0)^0$ is a collineation of Π_{V^*} to $\Pi_{V'^*}$. By Corollary 2.2, we have $n = n'$ and there is a semilinear isomorphism $u : V^* \to V'^*$ such that

$$f(P^0)^0 = u(P)$$

for every $P \in \mathcal{G}_1(V^*)$. Then

$$f(U) = u(U^0)^0$$

for every $U \in \mathcal{G}_{n-1}(V)$ and Proposition 1.11 implies that $f = (\check{u})_{n-1}$, where $\check{u} : V \to V'$ is the contragradient of u. $\qquad\square$

In what follows we will need some technical lemmas on mappings between projective spaces.

Lemma 2.1. *Let $\Pi = (\mathcal{P}, \mathcal{L})$ and $\Pi' = (\mathcal{P}', \mathcal{L}')$ be projective spaces of the same finite dimension. If $f : \mathcal{P} \to \mathcal{P}'$ is an injection transferring lines to lines then f is a collineation of Π to Π'.*

Proof. For any distinct points $p, q \in \mathcal{P}$ the line $f(p)f(q)$ coincides with the line $f(pq)$. This means that $f(\mathcal{P})$ is a subspace of Π' and f is a collineation of Π to the projective space $\Pi'' = (f(\mathcal{P}), f(\mathcal{L}))$. Then

$$\dim \Pi'' = \dim \Pi = \dim \Pi' < \infty$$

which implies that $f(\mathcal{P})$ coincides with $\mathcal{P}'$ (since in a projective space any two incident subspaces of the same finite dimension are coincident). $\qquad\square$

Lemma 2.2. *Let $\Pi = (\mathcal{P}, \mathcal{L})$ and $\Pi' = (\mathcal{P}', \mathcal{L}')$ be projective spaces. Suppose that $f : \mathcal{P} \to \mathcal{P}'$ satisfies the following condition: there is an injection $g : \mathcal{L} \to \mathcal{L}'$ such that for every line $l \in \mathcal{L}$ the image $f(l)$ is contained in the line $g(l)$. Then f is injective or constant.*

Proof. We assume that the mapping f is non-constant and show that it is injective. If $f(p) = f(q)$ for some distinct points $p, q \in \mathcal{P}$ then there is a point $t \in \mathcal{P}$ such that $f(t) \neq f(p)$ (otherwise, f is constant).

If $t \notin pq$ then the lines tp and tq are distinct. Their images $f(tp)$ and $f(tq)$ are contained in the lines $g(tp)$ and $g(tq)$, respectively. Each of these lines contains the points $f(t)$ and $f(p) = f(q)$. This means that the lines $g(tp)$ and $g(tq)$ are coincident. The latter is impossible, since g is injective.

In the case when $t \in pq$, we take any point $h \in \mathcal{P}$ which does not belong to the line pq. If $f(h) \neq f(p)$ then we apply the above arguments to the line pq and the point h. If $f(h) = f(p)$ then we consider the line ph and the point t. $\qquad\square$

Remark 2.1. Suppose that V and V' both are left 2-dimensional vector spaces over division rings. By Section 1.5, every semilinear isomorphism $l : V \to V'$ induces the bijection $(l)_1$ between the projective lines $\mathcal{G}_1(V)$ and $\mathcal{G}_1(V')$. Also, for every semilinear isomorphism $s : V \to V'^*$ the composition of $(s)_1$ and the annihilator mapping of $\mathcal{G}_1(V'^*)$ to $\mathcal{G}_1(V')$ is a bijection between $\mathcal{G}_1(V)$ and $\mathcal{G}_1(V')$. There are results in spirit of von Staudt's theorem characterizing the described above bijections of projective lines in terms of cross ratio and harmonic quadruples [Artin (1957), Section II.9] and [Baer (1952), Section III.4, Proposition 3]. More general statements concerning non-bijective mappings of projective lines were obtained in [Buekenhout (1965); Cojan (1985); James (1982); Klotzek (1988)]. These results can be considered as analogues of the Fundamental Theorem of Projective Geometry for projective lines.

2.3 Proof of Theorem 1.2

Let V and V' be left vector spaces over division rings R and R', respectively. Suppose that $\dim V = n$ is finite. For every mapping $l : V \to V'$ we denote by V_l the subspace of V' spanned by $l(V)$. Recall that $l : V \to V'$ is called a GL-*mapping* if it satisfies the following condition:

(GL) for every $u \in \mathrm{GL}(V)$ there is $u' \in \mathrm{GL}(V')$ such that $lu = u'l$.

Theorem 1.2 states that every non-trivial GL-mapping $l : V \to V'$ is a strong semilinear embedding if $n \geq 3$ and $\dim V_l \leq n$.

Let X be a finite subset of V containing more than one vector. Denote by $S(X)$ the group of all permutations on X. We say that X is a CL-*subset* if every permutation on X can be extended to a linear automorphism of

V. It is clear that every independent subset of V satisfies this condition. Since 0 is invariant for all linear automorphisms, every GL-subset does not contain 0.

Example 2.4. Let $x_1, \ldots, x_m$ be linearly independent vectors of V. Suppose that X is formed by $x_1, \ldots, x_m$ and the vector

$$x_{m+1} = -(x_1 + \cdots + x_m).$$

For every $i \in \{1, \ldots, m-1\}$ we take any linear automorphism $u_i \in \mathrm{GL}(V)$ such that

$$u_i(x_i) = x_{i+1}, \quad u_i(x_{i+1}) = x_i,$$

$$u_i(x_j) = x_j \text{ if } j \neq i, i+1, m+1.$$

It is clear that u_i transfers x_{m+1} to itself. Consider a linear automorphism $v \in \mathrm{GL}(V)$ leaving fixed every x_i for $i \leq m-1$ and transferring x_m to x_{m+1}. Then

$$v(x_{m+1}) = -(v(x_1) + \cdots + v(x_m))$$

$$= -(x_1 + \cdots + x_{m-1} - x_1 - \cdots - x_{m-1} - x_m) = x_m.$$

So, all transpositions $(x_i, x_{i+1}) \in S(X)$ can be extended to linear automorphisms of V. Since the group $S(X)$ is spanned by these transpositions, every permutation on X is extendable to a linear automorphism of V and X is a GL-subset.

To prove Theorem 1.2 we will use the following.

Proposition 2.1. *Every GL-subset $X \subset V$ is independent or it is formed by*

$$x_1, \ldots, x_m, -(x_1 + \cdots + x_m),$$

where $x_1, \ldots, x_m$ are linearly independent vectors.

Proof. Let $x_1, \ldots, x_k$ be the elements of X. Suppose that these vectors are not linearly independent and consider any maximal independent subset contained in X. We can assume that this subset is formed by $x_1, \ldots, x_m$, $m < k$. Then every x_p with $p > m$ is a linear combination of $x_1, \ldots, x_m$, i.e.

$$x_p = \sum_{l=1}^{m} a_l x_l.$$

Let $u \in \mathrm{GL}(V)$ be an extension of the transposition $(x_i, x_j) \in S(X)$ with $i, j \leq m$. Then

$$u(x_i) = x_j, \quad u(x_j) = x_i,$$

$$u(x_l) = x_l \text{ if } l \neq i, j.$$

We have

$$\sum_{l=1}^{m} a_l x_l = x_p = u(x_p) = \sum_{l=1}^{m} b_l x_l,$$

where

$$b_i = a_j, \; b_j = a_i \text{ and } b_l = a_l \text{ if } l \neq i, j.$$

Since $x_1, \ldots, x_m$ are linearly independent, the latter means that $a_i = a_j$. This equality holds for any pair $i, j \leq m$. Thus

$$x_p = a(x_1 + \cdots + x_m)$$

for some non-zero scalar $a \in R$. Let $v \in \mathrm{GL}(V)$ be an extension of the transposition $(x_1, x_p) \in S(X)$. Then

$$v(x_1) = x_p, \quad v(x_p) = x_1,$$

$$v(x_i) = x_i \text{ if } i \neq 1, p.$$

We have

$$x_1 = v(x_p) = a(v(x_1) + \cdots + v(x_m)) = a(x_p + x_2 + \cdots + x_m)$$

$$= a^2(x_1 + \cdots + x_m) + a(x_2 + \cdots + x_m) = a^2 x_1 + (a^2 + a)(x_2 + \cdots + x_m).$$

Hence $a^2 = 1$ and $a^2 + a = 0$ which implies that $a = -1$ and

$$x_p = -(x_1 + \cdots + x_m).$$

This equality holds for every $p > m$. Therefore, $k = m + 1$ and the second possibility is realized. $\qquad\qquad\square$

Let $l : V \to V'$ be a non-trivial GL-mapping, i.e. l satisfies the condition (GL) and the restriction of l to $V \setminus \{0\}$ is non-constant. Suppose also that $n \geq 2$.

Lemma 2.3. *The following assertions are fulfilled:*

(1) *for every $u \in \mathrm{GL}(V)$ there exists unique $\bar{u} \in \mathrm{GL}(V_l)$ such that $lu = \bar{u}l$,*
(2) *the mapping $u \to \bar{u}$ is a homomorphism of $\mathrm{GL}(V)$ to $\mathrm{GL}(V_l)$.*

Proof. (1). Let $u \in \mathrm{GL}(V)$. We take any $u' \in \mathrm{GL}(V')$ satisfying $lu = u'l$. For every $y \in l(V)$ there exists $x \in V$ such that $y = l(x)$ and we have

$$u'(y) = u'l(x) = lu(x) \in l(V).$$

Thus u' transfers $l(V)$ to itself. This means that the subspace V_l is invariant for u', since it is spanned by $l(V)$.

Consider another $u'' \in \mathrm{GL}(V')$ satisfying $lu = u''l$. If $y = l(x)$, $x \in V$ then

$$u'(y) = u'l(x) = lu(x) = u''l(x) = u''(y).$$

Therefore, $u'|_{l(V)} = u''|_{l(V)}$ which implies that $u'|_{V_l} = u''|_{V_l}$.

(2). If $u, v \in \mathrm{GL}(V)$ then

$$\overline{uv}\, l = luv = \overline{u}lv = \overline{u}\,\overline{v}l$$

and, by the statement (1), we have $\overline{u}\,\overline{v} = \overline{uv}$. $\qquad\square$

Lemma 2.4. *The following assertions are fulfilled:*

(1) $l(x) \neq 0$ *for every non-zero* $x \in V$,

(2) *if* $x, y \in V \setminus \{0\}$ *are linearly independent then* $l(x) \neq l(y)$.

Proof. (1). Suppose that $l(x) = 0$ for a certain non-zero vector $x \in V$. For every non-zero vector $y \in V$ there exists $u \in \mathrm{GL}(V)$ such that $y = u(x)$ and we have

$$l(y) = lu(x) = \overline{u}l(x) = 0.$$

This is impossible, since our GL-mapping is non-trivial.

(2). If $l(x) = l(y)$ for some linearly independent vectors $x, y \in V$ then

$$lu(x) = \overline{u}l(x) = \overline{u}l(y) = lu(y) \quad \forall\, u \in \mathrm{GL}(V).$$

Let $z \in V \setminus \{0\}$. In the case when x, z are linearly independent, we take any $u \in \mathrm{GL}(V)$ such that $u(x) = x$ and $u(y) = z$. Then

$$l(x) = lu(x) = lu(y) = l(z).$$

If z is a scalar multiple of x then y and z are linearly independent and the same arguments show that $l(y) = l(z)$. Thus $l|_{V \setminus \{0\}}$ is constant which contradicts the assumption that our GL-mapping is non-trivial. $\qquad\square$

For every subspace $S \subset V$ we denote by S_l the subspace of V' spanned by $l(S)$.

Lemma 2.5. *For every* $u \in \mathrm{GL}(V)$ *the following assertions are fulfilled:*

(1) *if a subspace $S \subset V$ is invariant for u then S_l is invariant for $\bar{u}$,*
(2) $(\mathrm{Ker}(\mathrm{id}_V - u))_l \subset \mathrm{Ker}(\mathrm{id}_{V_l} - \bar{u})$.

Proof. (1). An easy verification shows that $\bar{u}$ transfers $l(S)$ to itself. This means that the subspace S_l is invariant for $\bar{u}$, since it is spanned by $l(S)$.
(2). If $u(x) = x$ for a certain $x \in V$ then

$$\bar{u}l(x) = lu(x) = l(x)$$

and $l(x)$ belongs to $\mathrm{Ker}(\mathrm{id}_{V_l} - \bar{u})$. Hence

$$l(\mathrm{Ker}(\mathrm{id}_V - u)) \subset \mathrm{Ker}(\mathrm{id}_{V_l} - \bar{u})$$

which implies the required inclusion. □

Lemma 2.6. *If $\bar{u}$ is identity then u is identity or a homothety.*

Proof. If $\bar{u}$ is identity then

$$lu(x) = \bar{u}l(x) = l(x) \quad \forall\, x \in V.$$

If x and $u(x)$ are linearly independent then, by the second part of Lemma 2.4, $lu(x) \neq l(x)$ which contradicts the above equality. Thus $u(x)$ is a scalar multiple of x for every $x \in V$. This means that $(u)_1$ is the identity transformation of $\mathcal{G}_1(V)$ and Proposition 1.6 gives the claim. □

From this moment we suppose that $\dim V_l \leq n$ and use Proposition 2.1 to prove the following.

Lemma 2.7. *We have $\dim V_l = n$ and l transfers bases of V to bases of V_l.*

Proof. Let $B = \{x_1, \ldots, x_n\}$ be a base of V. Consider the GL-subset X formed by the vectors

$$x_1, \ldots, x_n \quad \text{and} \quad -(x_1 + \cdots + x_n).$$

Lemma 2.4 guarantees that $l|_X$ is a bijection to the subset $X' := l(X)$.
For every permutation s' on X' there is a permutation s on X such that $ls = s'l|_B$. Let $u \in \mathrm{GL}(V)$ be an extension of s. If $x' \in X'$ then $x' = l(x)$ for a certain $x \in X$ and

$$\bar{u}(x') = \bar{u}l(x) = lu(x) = ls(x) = s'l(x) = s'(x').$$

Thus $\bar{u}$ is an extension of s'. So, X' is a GL-subset of V_l.
By Proposition 2.1, X' is formed by $n + 1$ linearly independent vectors or it consists of

$$x'_1, \ldots, x'_n, -(x'_1 + \cdots + x'_n),$$

where $x_1', \ldots, x_n'$ are linearly independent. The condition $\dim V_l \leq n$ shows that the first possibility is not realized. Then $\dim V_l = n$ and every n-element subset of X' is a base of V_l. This means that $l(B)$ is a base of V_l, since it is contained in X'. $\qquad\square$

Lemma 2.8. *For every $(n-1)$-dimensional subspace $S \subset V$ the subspace S_l is $(n-1)$-dimensional.*

Proof. We take any $u \in GL(V)$ such that

$$\mathrm{Ker}(\mathrm{id}_V - u) = S. \tag{2.1}$$

By the second part of Lemma 2.5, we have

$$S_l \subset \mathrm{Ker}(\mathrm{id}_{V_l} - \overline{u}). \tag{2.2}$$

Every independent subset of V can be extended to a base of V and Lemma 2.7 guarantees that $\dim S_l \geq n - 1$. If $\dim S_l \geq n$ then

$$\dim(\mathrm{Ker}(\mathrm{id}_{V_l} - \overline{u})) \geq n$$

by (2.2). Since $\dim V_l = n$, the kernel of $\mathrm{id}_{V_l} - \overline{u}$ coincides with V_l and $\overline{u}$ is identity. If follows from Lemma 2.6 that u is identity or a homothety which contradicts (2.1). Therefore, S_l is $(n-1)$-dimensional. $\qquad\square$

Lemma 2.9. *For every non-zero subspace $S \subset V$ we have $\dim S_l = \dim S$.*

Proof. In the case when $n = 2$, the statement follows from the previous lemma. Suppose that $n \geq 3$ and consider any $(n-1)$-dimensional subspace $U \subset V$. If U is invariant for $u \in GL(V)$ then U_l is invariant for $\overline{u}$ (the first part of Lemma 2.5). This means that $l|_U$ is a GL-mapping of U to U_l. By Lemma 2.8, $\dim U_l = n - 1$ and this GL-mapping is non-trivial. We apply the above arguments to the GL-mapping $l|_U$ and establish that $\dim S_l = n-2$ for every $(n-2)$-dimensional subspace $S \subset U$. Step by step, we show that $\dim S_l = \dim S$ for every non-zero subspace $S \subset V$. $\qquad\square$

Lemma 2.10. $l(0) = 0$.

Proof. The condition $\dim V_l = n \geq 2$ implies the existence of 1-dimensional subspaces $P, Q \subset V$ such that P_l and Q_l are distinct 1-dimensional subspaces of V_l. Then $l(0) \in P_l \cap Q_l = 0$. $\qquad\square$

From this moment we suppose that $n \geq 3$. Consider the mapping f which transfers every $P \in \mathcal{G}_1(V)$ to $P_l \in \mathcal{G}_1(V_l)$. For any two linearly independent vectors $x, y \in V$ there is a base of V containing them and

Lemma 2.7 implies that $l(x)$ and $l(y)$ are linearly independent. This means that f is injective. The mapping f sends the line of Π_V defined by a 2-dimensional subspace $S \subset V$ to a subset in the line of Π_{V_l} corresponding to S_l. The projective space Π_{V_l} is spanned by $f(\mathcal{G}_1(V))$. Hence $f(\mathcal{G}_1(V))$ is not contained in a line of Π_{V_l}. Corollary 2.1 implies that f is induced by a semilinear injection $\tilde{l} : V \to V_l$. Then for every non-zero vector $x \in V$ there is a non-zero scalar $a_x \in R'$ such that

$$l(x) = a_x \tilde{l}(x).$$

It follows from Lemma 2.7 that $\tilde{l}$ transfers every base of V to a base of V_l. So, $\tilde{l}$ is a strong semilinear embedding of V in V_l.

Lemma 2.11. $a_x = a_y$ for all $x, y \in V \setminus \{0\}$.

Proof. Since $\tilde{l}$ is a GL-mapping, for every $u \in \mathrm{GL}(V)$ there exists $\tilde{u} \in \mathrm{GL}(V_l)$ such that $\tilde{l}u = \tilde{u}\tilde{l}$. Then

$$a_x \overline{u}\tilde{l}(x) = \overline{u}l(x) = lu(x) = a_{u(x)}\tilde{l}u(x) = a_{u(x)}\tilde{u}\tilde{l}(x)$$

and

$$\overline{u}\tilde{l}(x) = a_x^{-1}a_{u(x)}\tilde{u}\tilde{l}(x) \tag{2.3}$$

for every non-zero $x \in V$. The mappings $\overline{u}\tilde{l}$ and $\tilde{u}\tilde{l}$ both are semilinear injections and (2.3) guarantees that $(\overline{u}\tilde{l})_1 = (\tilde{u}\tilde{l})_1$. By Proposition 1.6, there exists a non-zero scalar $b_u \in R'$ such that

$$\overline{u}\tilde{l} = b_u \tilde{u}\tilde{l}.$$

Then (2.3) implies that

$$a_{u(x)} = a_x b_u.$$

Let $x, y \in V \setminus \{0\}$. We take any vector $z \in V$ which does not belong to the subspace spanned by x and y (since $n \geq 3$, this is possible). Consider any $u \in \mathrm{GL}(V)$ satisfying $u(x) = y$ and $u(z) = z$. Then $a_z = a_{u(z)} = a_z b_u$ which means that $b_u = 1$ and $a_y = a_{u(x)} = a_x b_u = a_x$. $\square$

Lemmas 2.10 and 2.11 show that l is a scalar multiple of the strong semilinear embedding $\tilde{l}$ which means that l is a strong semilinear embedding.

Problem 2.2. If $n = 2$ then l induces an injective mapping of the projective line $\mathcal{G}_1(V)$ to the projective line $\mathcal{G}_1(V_l)$. Is it possible to apply one of generalizations of von Staudt's theorem (Remark 2.1) to this mapping?

2.4 m-independent subsets in projective spaces

Let $\Pi = (\mathcal{P}, \mathcal{L})$ be a projective space. Recall that a subset $\mathcal{X} \subset \mathcal{P}$ is *independent* if the subspace spanned by $\mathcal{X}$ cannot be spanned by a proper subset of $\mathcal{X}$. Let m be a natural number not less than 2. We say that $\mathcal{X} \subset \mathcal{P}$ is an *m-independent* subset of Π if it contains at least m points and every m-element subset of $\mathcal{X}$ is independent. Note that any two distinct points form an independent subset and every subset containing more than one point is 2-independent.

An m-independent subset consisting of $m + 1$ points is called an m-*simplex* if it is not independent [Baer (1952), Section III.3]. By this definition, any triple of collinear points is a 2-simplex.

Let V be a left vector space over a division ring R. Suppose that $\dim V = n$ is finite and not less than 2. A subset of the associated projective space or the projective line is independent if and only if it consists of $P_1, \ldots, P_m \in \mathcal{G}_1(V)$ such that non-zero vectors $x_1 \in P_1, \ldots, x_m \in P_m$ are linearly independent. In the case when $n \geq 3$, the annihilator mapping induces a collineation between the projective spaces Π_V^* and Π_{V^*}; thus a subset of the dual projective space Π_V^* is independent if and only if it consists of $U_1, \ldots, U_m \in \mathcal{G}_{n-1}(V)$ such that

$$\dim(U_1 \cap \cdots \cap U_m) = n - m.$$

If $x_1, \ldots, x_m$ are linearly independent vectors of V and $a_1, \ldots, a_m \in R$ are non-zero scalars then

$$\langle x_1 \rangle, \ldots, \langle x_m \rangle \quad \text{and} \quad \langle a_1 x_1 + \cdots + a_m x_m \rangle$$

form an m-simplex.

Lemma 2.12. *If* $P_1, \ldots, P_{m+1} \in \mathcal{G}_1(V)$ *form an m-simplex then there exist non-zero vectors* $x_1 \in P_1, \ldots, x_m \in P_m$ *such that*

$$P_{m+1} = \langle x_1 + \cdots + x_m \rangle.$$

Proof. We take non-zero vectors $y_1 \in P_1, \ldots, y_m \in P_m$. Then

$$P_{m+1} = \langle a_1 y_1 + \cdots + a_m y_m \rangle$$

and every scalar a_i is non-zero. The vectors $x_i := a_i y_i$ are as required. $\square$

Remark 2.2. It is clear that $P_1, \ldots, P_l \in \mathcal{G}_1(V)$ form an m-independent subset if and only if non-zero vectors $x_1 \in P_1, \ldots, x_l \in P_l$ form an m-independent subset of V, i.e. any distinct $x_{j_1}, \ldots, x_{j_m}$ are linearly independent vectors.

In Section 1.5 we identify the groups $\mathrm{P\Gamma L}(V)$ and $\mathrm{PGL}(V)$ with the groups of all transformations of $\mathcal{G}_1(V)$ induced by semilinear automorphisms of V and linear automorphisms of V, respectively.

Proposition 2.2. *If $\{P_1, \ldots, P_{m+1}\}$ and $\{P'_1, \ldots, P'_{m+1}\}$ are m-simplices in $\mathcal{G}_1(V)$ then there is an element of $\mathrm{PGL}(V)$ transferring every P_i to P'_i. If $\{P_1, \ldots, P_{n+1}\}$ is an n-simplex in $\mathcal{G}_1(V)$ then the following two conditions are equivalent:*

(1) *the identity transformation of $\mathcal{G}_1(V)$ is the unique element of $\mathrm{PGL}(V)$ leaving fixed every P_i,*
(2) *R is a field.*

If R is a field then for any n-simplices $\{P_1, \ldots, P_{n+1}\}$ and $\{P'_1, \ldots, P'_{n+1}\}$ in $\mathcal{G}_1(V)$ there is the unique element of $\mathrm{PGL}(V)$ transferring every P_i to P'_i.

Proof. For m-simplices $\{P_1, \ldots, P_{m+1}\}$ and $\{P'_1, \ldots, P'_{m+1}\}$ we choose non-zero vectors

$$x_1 \in P_1, \ldots, x_m \in P_m \text{ and } x'_1 \in P'_1, \ldots, x'_m \in P'_m$$

such that

$$P_{m+1} = \langle x_1 + \cdots + x_m \rangle \text{ and } P'_{m+1} = \langle x'_1 + \cdots + x'_m \rangle.$$

If u is a linear automorphism of V transferring every x_i to x'_i then

$$u(x_1 + \cdots + x_m) = x'_1 + \cdots + x'_m$$

and $(u)_1$ is as required.

Now we consider an n-simplex $\{P_1, \ldots, P_{n+1}\}$ and choose non-zero vectors $x_1 \in P_1, \ldots, x_n \in P_n$ satisfying

$$P_{n+1} = \langle x_1 + \cdots + x_n \rangle.$$

(1) $\Longrightarrow$ (2). We take any non-zero scalar $a \in R$ and denote by u the linear automorphism of V sending every x_i to ax_i. Then $u(P_i) = P_i$ for all i and, by our hypothesis, $(u)_1$ is the identity transformation of $\mathcal{G}_1(V)$. It follows from Proposition 1.6 that u is a homothety. Then $u(x) = ax$ for all $x \in V$. Since u is linear, a belongs to the center of R (Example 1.15). So, every non-zero element of R belongs to the center, i.e. R is a field.

(2) $\Longrightarrow$ (1). Suppose that u is a linear automorphism of V satisfying $u(P_i) = P_i$ for all i. Then $u(x_i) = a_i x_i$ for every $i \in \{1, \ldots, n\}$. The equality $u(P_{m+1}) = P_{m+1}$ implies that $a_1 = a_2 = \cdots = a_n$. Since R is a field, u is a homothety and $(u)_1$ is the identity transformation of $\mathcal{G}_1(V)$. $\square$

Now we present examples of m-independent subsets containing more than $m + 1$ elements.

Example 2.5. Let $x_1, \ldots, x_m \in V$ be linearly independent vectors and let $a_1, \ldots, a_m$ be non-zero elements of R. Consider the subset of $\mathcal{G}_1(V)$ formed by

$$\langle x_1 \rangle, \ldots, \langle x_m \rangle, \ \langle x_1 + \cdots + x_m \rangle, \ \langle a_1 x_1 + \cdots + a_m x_m \rangle.$$

This subset is m-independent if and only if the vectors

$$x_i + x_j, \ a_i x_i + a_j x_j$$

are linearly independent for any pair of distinct i, j. The latter condition is equivalent to the fact that all scalars a_i are pairwise distinct. If R is a finite field then m-independent subsets of such type exist only in the case when $|R| \geq m + 1$.

Example 2.6. Suppose that $\mathcal{Y} \subset \mathcal{G}_1(V)$ is an m-simplex and $m < n$. We take any $P \in \mathcal{G}_1(V)$ which is not contained in the subspace spanned by $\mathcal{Y}$. Then $\mathcal{Y} \cup \{P\}$ is an m-independent subset. It contains independent subsets consisting of $m + 1$ elements, but $\mathcal{Y} \cup \{P\}$ is not $(m + 1)$-independent.

Lemma 2.13. *Let $\mathcal{X}$ be an m-independent subset of $\mathcal{G}_1(V)$ consisting of $m + 2$ elements. If $\mathcal{X}$ is not $(m + 1)$-independent then it is one of the subsets considered in Examples 2.5 and 2.6. In particular, every n-independent subset of $\mathcal{G}_1(V)$ consisting of $n + 2$ elements is the subset from Example 2.5 with $m = n$.*

Proof. There is a subset $\mathcal{Y} \subset \mathcal{X}$ which consists of $m + 1$ elements and is not independent. It is clear that $\mathcal{Y}$ is an m-simplex. Denote by $\mathcal{S}$ the $(m - 1)$-dimensional subspace of Π_V spanned by $\mathcal{Y}$. Consider the unique element $P \in \mathcal{X}$ which is not contained in $\mathcal{Y}$. If $P \in \mathcal{S}$ then $\mathcal{X}$ is the subset from Example 2.5. In the case when $P \notin \mathcal{S}$, we get the subset constructed in Example 2.6. $\square$

Proposition 2.3. *If the division ring R is infinite then for any integers m and l such that*

$$2 \leq m \leq n \ \text{and} \ m \leq l$$

there is an m-independent subset of $\mathcal{G}_1(V)$ consisting of l elements.

To prove Proposition 2.3 we need the following.

Lemma 2.14. *Let $\mathcal{Y}$ be a finite set whose elements are proper subspaces of V. If R is infinite then there exists $P \in \mathcal{G}_1(V)$ such that $P \not\subset S$ for every $S \in \mathcal{Y}$.*

Proof. Since every proper subspace of V is contained in an $(n-1)$-dimensional subspace, it is sufficient to consider the case when each element of $\mathcal{Y}$ is an $(n-1)$-dimensional subspace. We prove the statement by induction on n. The case when $n = 2$ is trivial and we suppose that $n \geq 3$. By inductive hypothesis, every $S \in \mathcal{Y}$ contains a 1-dimensional subspace P_S which is not contained in any other element of $\mathcal{Y}$. Let S and U be distinct elements of $\mathcal{Y}$ (the statement is trivial if $\mathcal{Y}$ is a one-element set). It is clear that P_S and P_U are distinct. Every element of $\mathcal{Y}$ intersects $P_S + P_U$ in a 1-dimensional subspace (otherwise it contains both P_S and P_U which is impossible). If R is infinite then $P_S + P_U$ contains infinitely many 1-dimensional subspaces and we get the claim. □

Proof of Proposition 2.3. The statement will be proved by induction on l. The case when $l = m$ is trivial. Suppose that $l > m$ and $\mathcal{X} \subset \mathcal{G}_1(V)$ is an m-independent subset consisting of $l - 1$ elements. Let $\mathcal{Y}$ be the set of all $(m-1)$-dimensional subspaces of type $P_1 + \cdots + P_{m-1}$, where $P_1, \ldots, P_{m-1}$ are distinct elements of $\mathcal{X}$. This set is finite and Lemma 2.14 implies the existence of $P \in \mathcal{G}_1(V)$ which is not contained in any element of $\mathcal{Y}$. Then $\mathcal{X} \cup \{P\}$ is an m-independent subset. □

Problem 2.3. In the case when R is a finite field, is it possible to determine the maximal integer $l(m)$ such that $\mathcal{G}_1(V)$ contains an m-independent subset consisting of $l(m)$ elements?

Let $k \in \{1, \ldots, n-1\}$. Consider the following equivalence relation for subsets of the Grassmannian $\mathcal{G}_k(V)$: two subsets are *equivalent* if there is an element of $\mathrm{P\Gamma L}(V)$ transferring one of them to the other. Also, we say that two subsets of $\mathcal{G}_k(V)$ are *linearly equivalent* if there exists an element of $\mathrm{PGL}(V)$ sending one of them to the other. Recall that we identify the groups $\mathrm{P\Gamma L}(V)$ and $\mathrm{PGL}(V)$ with the groups of all transformations of $\mathcal{G}_k(V)$ induced by semilinear automorphisms of V and linear automorphisms of V, respectively.

Lemma 2.15. *Subsets $\mathcal{X}, \mathcal{Y} \subset \mathcal{G}_k(V)$ are equivalent or linearly equivalent if and only if the subsets $\mathcal{X}^0, \mathcal{Y}^0 \subset \mathcal{G}_{n-k}(V^*)$ are equivalent or linearly equivalent, respectively.*

Proof. Let u be a semilinear automorphism of V. Then $(u)_k$ transfers $\mathcal{X}$ to $\mathcal{Y}$ if and only if $(\check{u})_{n-k}$ transfers $\mathcal{X}^0$ to $\mathcal{Y}^0$.   □

In the case when $k = 1, n-1$, independent subsets of $\mathcal{G}_k(V)$ containing the same number of elements are linearly equivalent and Proposition 2.2 implies that any two m-simplices of $\mathcal{G}_k(V)$ are linearly equivalent.

The subsets considered in Examples 2.5 and 2.6 are not equivalent. Any two m-independent subsets from Example 2.6 are linearly equivalent. For m-independent subsets from Example 2.5 this fails. In some cases there are infinitely many equivalence classes for subsets of such type.

Example 2.7. Suppose that R is a field such that the groups $\mathrm{P\Gamma L}(V)$ and $\mathrm{PGL}(V)$ are coincident, see Remark 1.6. The latter condition holds, for example, if R is the field of rational, real or p-adic numbers or the Galois field $\mathrm{GF}(p)$, where $p > 1$ is a prime number. Let $x_1, \dots, x_n$ be a base of V. Consider the set A formed by all vectors

$$a = (a_1, \dots, a_{n-1}) \in R^{n-1}$$

such that the subset $\mathcal{X}_a \subset \mathcal{G}_1(V)$ consisting of

$$P_1 = \langle x_1 \rangle, \dots, P_n = \langle x_n \rangle, \ P_{n+1} = \langle x_1 + \dots + x_n \rangle$$

and

$$\langle a_1 x_1 + \dots + a_{n-1} x_{n-1} + x_n \rangle$$

is n-independent, see Example 2.5. The set A is formed by all vectors of R^{n-1} whose coordinates are pairwise distinct and different from 0 and 1. If $a, b \in A$ are distinct then the same holds for the subsets $\mathcal{X}_a$ and $\mathcal{X}_b$. Suppose that $\mathcal{X}_a$ and $\mathcal{X}_b$ are equivalent and f is an element of $\mathrm{PGL}(V)$ transferring $\mathcal{X}_a$ to $\mathcal{X}_b$. Then we have

$$f(\mathcal{Y}) = \{P_1, \dots, P_{n+1}\}$$

for a certain $(n+1)$-element subset $\mathcal{Y} \subset \mathcal{X}_a$. Every $(n+1)$-element subset of $\mathcal{X}_a$ is an n-simplex and, by Proposition 2.2, f is completely defined by the restriction $f|_\mathcal{Y}$. Thus there is only a finite number of elements of $\mathrm{PGL}(V)$ which transfer $(n+1)$-element subsets of $\mathcal{X}_a$ to $\{P_1, \dots, P_{n+1}\}$. This number is not greater than $(n+2)!$. Indeed, there are precisely $n+2$ distinct $(n+1)$-element subsets of $\mathcal{X}_a$ and each of these subsets admits precisely $(n+1)!$ permutations. This means that for every $a \in A$ there is only a finite number of $b \in A$ such that $\mathcal{X}_a$ and $\mathcal{X}_b$ are equivalent. Therefore, if R is infinite then there are infinitely many equivalence classes of n-independent subsets of $\mathcal{G}_1(V)$ consisting of $n+2$ elements.

Remark 2.3. Suppose that A is the set from the previous example and $R = \mathbb{R}$. Consider the following equivalence relation $\sim$ on the set A: $a \sim b$ if $\mathcal{X}_a$ and $\mathcal{X}_b$ are equivalent. The corresponding quotient space $A/{\sim}$ is an $(n-1)$-dimensional manifold.

2.5 PGL-subsets

As in the previous section, we consider a left vector space V over a division ring R and suppose that $\dim V = n$ is finite and not less than 2. Let $k \in \{1, \ldots, n-1\}$. We say that $\mathcal{X} \subset \mathcal{G}_k(V)$ is a PGL-*subset* if every permutation on $\mathcal{X}$ can be extended to an element of $\mathrm{PGL}(V)$. The group of all permutations on $\mathcal{X}$ will be denoted by $S(\mathcal{X})$.

Lemma 2.16. *A subset* $\mathcal{X} \subset \mathcal{G}_k(V)$ *is a* PGL-*subset if and only if* $\mathcal{X}^0$ *is a* PGL-*subset of* $\mathcal{G}_{n-k}(V^*)$.

Proof. For every permutation $s \in S(\mathcal{X})$ the mapping $\check{s}$ transferring every $U \in \mathcal{X}^0$ to $s(U^0)^0$ is a permutation on $\mathcal{X}^0$ and $s \to \check{s}$ is an isomorphism between the groups $S(\mathcal{X})$ and $S(\mathcal{X}^0)$. Let u be a linear automorphism of V. Then $(u)_k$ is an extension of $s \in S(\mathcal{X})$ if and only if $(\check{u})_{n-k}$ is an extension of $\check{s}$. $\square$

In this section we investigate PGL-subsets of projective spaces and lines. By Lemma 2.16, we can consider only PGL-subsets of $\mathcal{G}_1(V)$. All independent subsets of Π_V are PGL-subsets and it follows from Proposition 2.2 that every m-simplex is a PGL-subset.

Example 2.8. Let P_1, P_2, P_3, P_4 be a *harmonic quadruple* in $\mathcal{G}_1(V)$, i.e. the characteristic of R is not equal to 2 and there are linearly independent vectors $x_1, x_2 \in V$ such that

$$P_1 = \langle x_1 \rangle, \ P_2 = \langle x_2 \rangle, \ P_3 = \langle x_1 + x_2 \rangle, \ P_4 = \langle x_1 - x_2 \rangle.$$

See [Baer (1952), Section III.4, Remark 5]. Note that in harmonic quadruples the ordering of points is essential. Now we suppose that the characteristic of R is equal to 3 and consider the permutation group $S(\mathcal{X})$, where $\mathcal{X}$ is the set formed by all P_i. Let u, v, w be linear automorphisms of V satisfying the following conditions

$$
\begin{aligned}
u(x_1) &= x_2, & u(x_2) &= x_1, \\
v(x_1) &= -x_1, & v(x_2) &= x_1 + x_2, \\
w(x_1) &= x_1, & w(x_2) &= -x_2.
\end{aligned}
$$

Since the characteristic of R is equal to 3, we have

$$v(x_1 - x_2) = -x_1 - (x_1 + x_2) = -2x_1 - x_2 = x_1 - x_2.$$

An easy verification shows that the restrictions of $(u)_1, (v)_1, (w)_1$ to $\mathcal{X}$ are the transpositions

$$(P_1, P_2), (P_2, P_3), (P_3, P_4) \in S(\mathcal{X}),$$

respectively. Since $S(\mathcal{X})$ is spanned by these transpositions, $\mathcal{X}$ is a PGL-subset. This implies that

$$P_{\delta(1)}, \ P_{\delta(2)}, \ P_{\delta(3)}, \ P_{\delta(4)}$$

is a harmonic quadruple for every permutation δ on the set $\{1, 2, 3, 4\}$. Indeed, if $u \in \mathrm{GL}(V)$ and $(u)_1$ is an extension of the permutation on $\mathcal{X}$ corresponding to δ then

$$P_{\delta(1)} = \langle u(x_1) \rangle, \ P_{\delta(2)} = \langle u(x_2) \rangle,$$

$$P_{\delta(3)} = \langle u(x_1) + u(x_2) \rangle, \ P_{\delta(4)} = \langle u(x_1) - u(x_2) \rangle.$$

Therefore, the ordering of points is not essential in our case and we can say that $\mathcal{X}$ is a *harmonic subset*. Any two harmonic subsets of $\mathcal{G}_1(V)$ are linearly equivalent.

Now we show that every PGL-subset is m-independent for a certain number m.

Lemma 2.17. *If $\mathcal{X}$ is a PGL-subset of $\mathcal{G}_1(V)$ then there is a number m such that $\mathcal{X}$ is an m-independent subset and all elements of $\mathcal{X}$ are contained in a certain m-dimensional subspace of V.*

Proof. Let us take any maximal independent subset $\mathcal{Y}$ contained in $\mathcal{X}$. Suppose that $|\mathcal{Y}| = m$. For every m-element subset $\mathcal{Z} \subset \mathcal{X}$ there is a permutation on $\mathcal{X}$ transferring $\mathcal{Y}$ to $\mathcal{Z}$. Since this permutation is extendable to an element of $\mathrm{PGL}(V)$, the subset $\mathcal{Z}$ is independent. Every element of $\mathcal{X}$ is contained in the m-dimensional subspace of V spanned by the elements of $\mathcal{Y}$ (otherwise, $\mathcal{Y}$ is not a maximal independent subset contained in $\mathcal{X}$). $\square$

We classify PGL-subsets of $\mathcal{G}_1(V)$ under the assumption that R is a field. For the non-commutative case the classification problem is open.

Proposition 2.4 ([Pankov (2013)]). *Suppose that R is a field. Then for every PGL-subset $\mathcal{X} \subset \mathcal{G}_1(V)$ one of the following possibilities is realized:*

- $\mathcal{X}$ is an independent subset,
- $\mathcal{X}$ is an m-simplex with $m \in \{2, \ldots, n\}$,
- the characteristic of R is equal to 3 and $\mathcal{X}$ is a harmonic subset.

To prove Proposition 2.4 we use the following simple lemma.

Lemma 2.18. *If R is a field and $f \in \mathrm{PGL}(V)$ transfers $P \in \mathcal{G}_1(V)$ to $Q \in \mathcal{G}_1(V)$ then for any non-zero vectors $x \in P$ and $y \in Q$ there exists $u \in \mathrm{GL}(V)$ such that $f = (u)_1$ and $u(x) = y$.*

Proof. If $v \in \mathrm{GL}(V)$ and $f = (v)_1$ then $v(x) = ay$ and the linear automorphism $u := a^{-1}v$ is as required. $\square$

Remark 2.4. The latter statement fails if R is non-commutative. In this case, a scalar multiple of a non-zero linear transformation of V is linear if and only if the scalar belongs to the center of R.

Proof of Proposition 2.4. Let $P_1, \ldots, P_k$ be the elements of $\mathcal{X}$. In the case when $\mathcal{X}$ is not independent, we take any maximal independent subset contained in $\mathcal{X}$. Suppose that this subset is formed by $P_1, \ldots, P_m$. It follows from Lemma 2.17 that $\mathcal{X}$ is m-independent. Then $\mathcal{X}$ is an m-simplex if $k = m + 1$.

Consider the case when $k \geq m + 2$. If $p \geq m + 2$ then $P_1, \ldots, P_{m+1}, P_p$ form an m-independent subset of the type considered in Example 2.5. We take non-zero vectors $x_1 \in P_1, \ldots, x_m \in P_m$ such that

$$x_{m+1} := x_1 + \cdots + x_m \in P_{m+1}$$

and $P_p = \langle x_p \rangle$, where

$$x_p = b_1 x_1 + \cdots + b_m x_m,$$

$b_1, \ldots, b_m$ are non-zero and mutually distinct. Let v be a linear automorphism of V such that $(v)_1$ is an extension of the transposition

$$(P_{m+1}, P_p) \in S(\mathcal{X}).$$

By Lemma 2.18, we can suppose that v sends x_{m+1} to x_p. Since $x_1, \ldots, x_m$ are linearly independent and $v(P_i) = P_i$ for every $i \leq m$, the equality

$$v(x_1) + \cdots + v(x_m) = v(x_{m+1}) = x_p = b_1 x_1 + \cdots + b_m x_m$$

shows that $v(x_i) = b_i x_i$ for every $i \leq m$. Then

$$v(x_p) = b_1 v(x_1) + \cdots + b_m v(x_m) = b_1^2 x_1 + \cdots + b_m^2 x_m \in P_{m+1}$$

which means that

$$b_1^2 = b_2^2 = \cdots = b_m^2.$$

Thus $0 = b_i^2 - b_j^2 = (b_i + b_j)(b_i - b_j)$ and $b_i = \pm b_j$ for any $i, j \leq m$. Since $b_1, \ldots, b_m$ are mutually distinct, we have $m = 2$ and $P_p = \langle x_1 - x_2 \rangle$ for every $p \geq 4$. This means that the subset $\mathcal{X}$ is formed by

$$P_1 = \langle x_1 \rangle, \; P_2 = \langle x_2 \rangle, \; P_3 = \langle x_1 + x_2 \rangle, \; P_4 = \langle x_1 - x_2 \rangle,$$

i.e. P_1, P_2, P_3, P_4 is a harmonic quadruple.

We choose a linear automorphism w of V such that $(w)_1$ is an extension of the transposition

$$(P_1, P_3) \in S(\mathcal{X})$$

and $w(x_1) = x_1 + x_2$ (Lemma 2.18). Since $w(P_2) = P_2$ and $w(P_3) = P_1$, we have

$$w(x_1 + x_2) = w(x_1) + w(x_2) = (x_1 + x_2) + cx_2 \in P_1.$$

Then $c = -1$ and $w(x_2) = -x_2$. The equality $w(P_4) = P_4$ implies that

$$w(x_1 - x_2) = w(x_1) - w(x_2) = (x_1 + x_2) + x_2 = x_1 + 2x_2 \in P_4.$$

Thus $x_1 + 2x_2 = x_1 - x_2$ which means that $2 = -1$ and the characteristic of R is equal to 3. $\qquad\square$

Problem 2.4. It will be interesting to obtain some information concerning PGL-subsets of $\mathcal{G}_k(V)$ in the case when $1 < k < n$. Also, we do not consider so-called PΓL-*subsets*. A subset $\mathcal{X} \subset \mathcal{G}_k(V)$ belongs to this class if every permutation on $\mathcal{X}$ can be extended to an element of PΓL(V).

2.6 Generalized MacWilliams theorem

As above, we suppose that V is a left vector space over a division ring R and $\dim V = n$ is finite and not less than 2. We fix a base $e_1, \ldots, e_n$ of V and denote by $e_1^*, \ldots, e_n^*$ the vectors in the dual base of V^* such that

$$e_i^* \cdot e_j = \delta_{ij}.$$

A semilinear automorphism of V is called *monomial* if it transfers every e_i to a scalar multiple of e_j. So, if l is a σ-linear monomial automorphism of V then

$$l(a_1 e_1 + \cdots + a_n e_n) = \sigma(a_1) b_1 e_{\delta(1)} + \cdots + \sigma(a_n) b_n e_{\delta(n)},$$

where $b_1, \ldots, b_n$ are non-zero scalars and δ is a permutation on $\{1, \ldots, n\}$.

We will need the following lemma which states that the contragradient of a monomial automorphism is monomial.

Lemma 2.19. *If $s \in \Gamma L(V)$ transfers every e_i to a scalar multiple of $e_{\delta(i)}$, where δ is a permutation on $\{1, \ldots, n\}$, then the contragradient $\check{s} \in \Gamma L(V^*)$ sends every e_i^* to a scalar multiple of $e_{\delta(i)}^*$.*

Proof. For any $i, j \in \{1, \ldots, n\}$ we have

$$s^*(e_{\delta(i)}^*) \cdot e_j = \sigma^{-1}(e_{\delta(i)}^* \cdot s(e_j)) = \sigma^{-1}(e_{\delta(i)}^* \cdot a_j e_{\delta(j)}), \qquad (2.4)$$

where σ is the automorphism of R associated to s and every a_j is non-zero. We observe that (2.4) is non-zero only in the case when $i = j$. This means that $s^*(e_{\delta(i)}^*)$ is a scalar multiple of e_i^*. Then $\check{s} = (s^*)^{-1}$ sends e_i^* to a scalar multiple of $e_{\delta(i)}^*$. $\qquad \square$

Let C be a k-dimensional subspace of V such that the restriction of every e_i^* to C is non-zero. So,

$$e_1^*|_C, \ldots, e_n^*|_C$$

are non-zero vectors of C^* and we denote by $P_1, \ldots, P_n$ the corresponding elements of $\mathcal{G}_1(C^*)$. Note that these elements are not necessarily distinct. In the case when $k = 1$, they are coincident. If $k \geq 2$ then $P_1, \ldots, P_n$ span the projective space or the projective line associated to C^* (since every vector of C^* is a linear combination of $e_1^*|_C, \ldots, e_n^*|_C$). The collection $P_1, \ldots, P_n$ is called the *projective system* associated to C and denoted by $\mathcal{P}(C)$ [Tsfasman, Vlăduţ and Nogin (2007), Section 1.1].

Let C' be another one k-dimensional subspace of V satisfying the same condition. The 1-dimensional subspace containing $e_i^*|_{C'}$ will be denoted by P_i'. Then $P_1', \ldots, P_n'$ form the projective system $\mathcal{P}(C')$.

Lemma 2.20. *For every semilinear isomorphism $u : C \to C'$ the following two conditions are equivalent:*

(1) *u is extendable to a semilinear monomial automorphism of V,*
(2) *the contragradient $\check{u} : C^* \to C'^*$ transfers $\mathcal{P}(C)$ to $\mathcal{P}(C')$.*

Proof. (1) $\Longrightarrow$ (2). Suppose that s is a semilinear monomial automorphism of V whose restriction to C coincides with u. If s maps every e_i to a scalar multiple of $e_{\delta(i)}$ for a certain permutation δ on $\{1, \ldots, n\}$ then, by Lemma 2.19, the contragradient $\check{s}$ sends every e_i^* to a scalar multiple of

$e^*_{\delta(i)}$. The equality $s(C) = C'$ implies that $\check{s}$ transfers C^0 to C'^0 (Proposition 1.11). Since V^*/C^0 and V^*/C'^0 can be naturally identified with C^* and C'^* (respectively), $\check{s}$ induces a semilinear isomorphism $\tilde{s} : C^* \to C'^*$ (Lemma 1.3). This semilinear isomorphism maps $e^*_i|_C$ to a scalar multiple of $e^*_{\delta(i)}|_{C'}$ and we have $\tilde{s}(P_i) = P'_{\delta(i)}$ for every i. So, we need to show that $\tilde{s}$ coincides with $\check{u}$.

Every vector of C^* can be presented in the form $x^*|_C$, where $x^* \in V^*$. It is clear that

$$\tilde{s}(x^*|_C) = \check{s}(x^*)|_{C'}.$$

If σ is the automorphism of R associated to u then for every $x \in C'$ we have

$$\tilde{s}(x^*|_C) \cdot x = \check{s}(x^*) \cdot x = (s^{-1})^*(x^*) \cdot x$$

$$= \sigma(x^* \cdot s^{-1}(x)) = \sigma(x^* \cdot u^{-1}(x)) = (u^{-1})^*(x^*|_C) \cdot x = \check{u}(x^*|_C) \cdot x$$

which gives the claim.

(2) $\Longrightarrow$ (1). Suppose that $\check{u}$ sends every P_i to $P'_{\delta(i)}$ for a certain permutation δ on $\{1, \ldots, n\}$. Then it transfers $e^*_i|_C$ to $a_i e^*_{\delta(i)}|_{C'}$, where a_i is a non-zero scalar. Let σ be the automorphism of R associated to u. Denote by v the σ-linear monomial automorphism of V^* which maps every e^*_i to $a_i e^*_{\delta(i)}$. Then

$$\check{u}(x^*|_C) = v(x^*)|_{C'}$$

for every $x^* \in V^*$. Lemma 2.19 guarantees that $\check{v}$ sends every e_i to a scalar multiple of $e_{\delta(i)}$, i.e. $\check{v}$ is monomial. If $x \in C$ then

$$x^* \cdot \check{v}(x) = x^* \cdot (v^{-1})^*(x) = \sigma(v^{-1}(x^*) \cdot x)$$

$$= \sigma(\check{u}^{-1}(x^*|_{C'}) \cdot x) = \sigma(u^*(x^*|_{C'}) \cdot x) = \sigma\sigma^{-1}(x^* \cdot u(x)) = x^* \cdot u(x).$$

Thus $\check{v}|_C$ coincides with u and $\check{v}$ is the required extension of u. $\square$

Remark 2.5. Suppose that $k = 1$. Then $P_i = C^*$ and $P'_i = C'^*$ for all i. Thus for every semilinear isomorphism $u : C \to C'$ the contragradient $\check{u}$ transfers $\mathcal{P}(C)$ to $\mathcal{P}(C')$. It follows from Lemma 2.20 that every semilinear isomorphism of C to C' can be extended to a semilinear monomial automorphism of V.

For every vector $x \in V$ we define the *weight* $||x||$ as the number of all i such that $e_i^* \cdot x$ is non-zero, i.e. the weight is the number of non-zero coordinates of x. Every semilinear monomial automorphism of V is weight preserving.

We prove the following version of the well-known MacWilliams theorem [MacWilliams (1961)].

Theorem 2.3. *Every weight preserving semilinear isomorphism* $u \colon C \to C'$, *i.e. such that*

$$||u(x)|| = ||x|| \quad \forall\, x \in C,$$

can be extended to a semilinear monomial automorphism of V.

For every subspace $S \subset V$ the *weight* $||S||$ is the number of all i such that $e_i^*|_S$ is non-zero. If P is a 1-dimensional subspace of V then $||P|| = ||x||$ for every non-zero vector $x \in P$.

Lemma 2.21. *Suppose that R is the finite field consisting of q elements. If S is an s-dimensional subspace of V then*

$$||S|| = \frac{1}{q^{s-1}} \sum_{P \in \mathcal{G}_1(S)} ||P||.$$

Proof. Recall that the number of all 1-dimensional subspaces contained in an m-dimensional subspace of V is equal to

$$\frac{q^m - 1}{q - 1}.$$

If $e_i^*|_S$ is non-zero then there are precisely

$$\frac{q^s - 1}{q - 1} - \frac{q^{s-1} - 1}{q - 1} = q^{s-1}$$

elements $P \in \mathcal{G}_1(S)$ such that $e_i^*|_P$ is non-zero (the number of all 1-dimensional subspaces in S minus the number of all 1-dimensional subspaces in the kernel of $e_i^*|_S$). This means that there are precisely

$$\frac{1}{q^{s-1}} \sum_{P \in \mathcal{G}_1(S)} ||P||$$

distinct i such that $e_i^*|_S$ is non-zero. $\qquad \square$

Proof of Theorem 2.3. The case when $k = 1$ is trivial (see Remark 2.5) and we suppose that $k \geq 2$.

By Lemma 2.20, it is sufficient to show that the contragradient $\check{u}$ transfers every P_i to a certain P'_j. For every $i \in \{1, \ldots, n\}$ we denote by S_i and S'_i the annihilators of P_i and P'_i, respectively. These are $(k-1)$-dimensional subspaces in C and C', respectively. We need to prove that u sends every S_i to a certain S'_j. Then, by Proposition 1.11, $\check{u}(P_i) = P'_j$ and we get the claim.

For every subspace $S \subset C$ the weight $||S||$ is equal to the number of all i such that P_i is not contained in S^0. If S is $(k-1)$-dimensional then S^0 is 1-dimensional. In this case, we have $||S|| \leq n - 1$ if and only if S coincides with a certain S_i (since $P_1, \ldots, P_n$ are not necessarily distinct, the inequality $||S|| < n - 1$ is possible). Similarly, for a $(k-1)$-dimensional subspace $S' \subset C'$ we have $||S'|| \leq n - 1$ if and only if S' coincides with a certain S'_i.

Since u is weight preserving, we have $||u(S)|| = ||S||$ for every 1-dimensional subspace $S \subset C$. If R is a finite field then Lemma 2.21 implies that the latter equality holds for every subspace $S \subset C$. Thus u transfers every S_i to a certain S'_j.

Consider the case when R is infinite. Suppose that there exists i such that $u(S_i) \neq S'_j$ for all j. Then the intersection of every S'_j with $u(S_i)$ is $(k-2)$-dimensional. Lemma 2.14 implies the existence of a vector $x' \in u(S_i)$ which does not belong to any S'_j. We have $||x'|| = n$ and $||u^{-1}(x')|| = n$. The latter contradicts the fact that $||S_i|| \leq n - 1$. □

Remark 2.6. In contrast to other presentations of this result, we do not assume that R is a finite field. The idea of the above proof is taken from [Ghorpade and Kaipa (2013)]. Other proofs of MacWilliams theorem can be found in [Bogart, Goldberg and Gordon (1978); Ward and Wood (1996)]. We refer [Fan, Liu and Puig (2003); Liu, Wu, Luo and Chen (2011); Liu and Zeng (2013)] for some generalizations of this result.

2.7 Linear codes

Let F be the finite field consisting of q elements. In other words, $F = \mathrm{GF}(q)$, where $q = p^r$ and $p > 1$ is a prime number. Consider the n-dimensional vector space $V = F^n$ with $n \geq 2$. Let $e_1, \ldots, e_n$ be the standard base of V, i.e.

$$e_1 = (1, 0, \ldots, 0), \ldots, e_n = (0, \ldots, 0, 1).$$

The dual base of V^* is formed by $e_1^*, \ldots, e_n^*$, where each e_i^* is the i-th coordinate functional

$$(a_1, \ldots, a_n) \to a_i.$$

If $r > 1$ then there are non-identity automorphisms of F and there exist semilinear transformations of V which are not linear.

In coding theory, a *linear* $[n, k]_q$ *code* is a k-dimensional subspace $C \subset V$ with $k < n$. The code is *non-degenerate* if the restriction of every e_i^* to C is non-zero. In what follows we will consider only non-degenerate linear codes.

Two linear $[n, k]_q$ codes are *equivalent* if there is a semilinear monomial automorphism of V transferring one of them to the other. In the case when this monomial automorphism is linear, we say that the codes are *linearly equivalent*. These equivalence relations are coincident only in the case when $r = 1$.

Example 2.9. Every linear $[n, 1]_q$ code consists of all vectors of type

$$(a_1 c, \ldots, a_n c),$$

where $a_1, \ldots, a_n$ are fixed non-zero elements of F. Any two linear $[n, 1]_q$ codes are linearly equivalent.

By Lemma 2.20, two linear $[n, k]_q$ codes C and C' are equivalent if and only if there is a semilinear isomorphism of C^* to C'^* transferring the projective system $\mathcal{P}(C)$ to the projective system $\mathcal{P}(C')$. These codes are linearly equivalent if and only if there is a linear isomorphism of C^* to C'^* satisfying the same condition. Also, Theorem 2.3 can be reformulated in terms of isomorphisms of linear codes.

An *automorphism* of a linear $[n, k]_q$ code C is a semilinear automorphism u of C which can be extended to a semilinear monomial automorphism of V. The latter is possible only in the case when the contragradient $\check{u}$ preserves $\mathcal{P}(C)$ (Lemma 2.20).

By [Tsfasman, Vlăduţ and Nogin (2007), Section 1.1], every collection of points in the projective space over a finite field defines a certain equivalence class of linear codes. Consider the k dimensional vector space $W = F^k$ such that $1 < k < n$. Suppose that $\mathcal{P}$ is a collection formed by certain $P_1, \ldots, P_n \in \mathcal{G}_1(W)$. We do not require that the elements of $\mathcal{P}$ are mutually distinct, i.e. the equality $P_i = P_j$ is possible for some pairs i, j, but we assume that the projective space or the projective line associated to W is spanned by the elements of $\mathcal{P}$. We choose non-zero vectors

$$x_1 \in P_1, \ldots, x_n \in P_n$$

and define the linear mapping $u : W^* \to V$ as follows

$$u(x^*) = (x^* \cdot x_1, \ldots, x^* \cdot x_n)$$

for every $x^* \in W^*$. By our assumption, W is spanned by the vectors $x_1, \ldots, x_n$ and there is no non-zero $x^* \in W^*$ such that $x^* \cdot x_i = 0$ for all i. Thus the mapping u is injective and $C = u(W^*)$ is a non-degenerate linear $[n, k]_q$ code.

Lemma 2.22. *Consider the mapping u as a linear isomorphism of W^* to C. Then the contragradient $\breve{u} : W \to C^*$ sends every P_i to the 1-dimensional subspace of C^* containing $e_i^*|_C$, in other words, $\mathcal{P}$ goes to the projective system $\mathcal{P}(C)$.*

Proof. We have

$$x^* \cdot x_i = e_i^* \cdot u(x^*) = x^* \cdot u^*(e_i^*|_C)$$

for every $x^* \in W^*$. This implies that $x_i = u^*(e_i^*|_C)$ and $\breve{u}(x_i) = e_i^*|_C$. $\square$

Let us take other non-zero vectors on $P_1, \ldots, P_n$ and define the corresponding linear mapping $s : W^* \to V$. We state that the codes C and $S = s(W^*)$ are linearly equivalent.

Consider u and s as linear isomorphisms of W^* to C and S, respectively. By Lemma 2.22, the composition of

$$u^* : C^* \to W \quad \text{and} \quad \dot{s} : W \to S^*$$

transfers the projective system $\mathcal{P}(C)$ to the projective system $\mathcal{P}(S)$. It follows from Lemma 2.20 that the contragradient of the composition is a linear isomorphism of C to S which can be extended to a linear monomial automorphism of V. This gives the claim.

Similarly, we show that for every permutation δ on $\{1, \ldots, n\}$ the linear $[n, k]_q$ codes obtained from $P_{\delta(1)}, \ldots, P_{\delta(n)}$ are linearly equivalent to C.

Now we take other collection $\mathcal{P}'$ consisting of $P_1', \ldots, P_n' \in \mathcal{G}_1(W)$ spanning the projective space or the projective line associated to W. We choose non-zero vectors on the elements of $\mathcal{P}'$ and consider the corresponding linear mapping $u' : W^* \to V$. Using Lemmas 2.20 and 2.22, we establish that the codes C and $C' = u'(W^*)$ are equivalent if and only if there is a semilinear automorphism of W transferring $\mathcal{P}$ to $\mathcal{P}'$. The codes are linearly equivalent if and only if there is a linear automorphism of W satisfying the same condition.

Example 2.10. Suppose that $n = k + 2$. Let $x_1, \ldots, x_k$ be a base of W. Consider the k-independent subset of $\mathcal{G}_1(W)$ formed by

$$\langle x_1 \rangle, \ldots, \langle x_k \rangle, \ \langle x_1 + \cdots + x_k \rangle, \ \langle a_1 x_1 + \cdots + a_k x_k \rangle,$$

where $a_1, \ldots, a_k$ are non-zero mutually distinct elements of the field F. This subset defines the equivalence class of linear $[k+2, k]_q$ codes containing the code formed by all vectors of type

$$(c_1, \ldots, c_k, c_1 + \cdots + c_k, a_1 c_1 + \cdots + a_k c_k).$$

Similarly, we can construct the equivalence class of linear $[k+1, k]_q$ codes corresponding to k-simplices of $\mathcal{G}_1(W)$. We refer [Tsfasman, Vlăduţ and Nogin (2007), Section 1.2] for various examples of linear codes.

Let C be a linear $[n, k]_q$ code and $1 < k < n$. An important parameter of the code C is the *minimal weight*

$$\min\{\|x\| : x \in C \setminus \{0\}\}.$$

It is clear that equivalent codes have the same minimal weight.

Example 2.11. The minimal weight of the linear code considered in Example 2.10 is equal to 3. The minimal weight of the linear $[k+1, k]_q$ codes defined by k-simplices of $\mathcal{G}_1(W)$ is equal to 2.

There is a simple geometric interpretation of the minimal weight.

Lemma 2.23. *The minimal weight of C is equal to n minus the maximal number of elements from the projective system $\mathcal{P}(C)$ contained in a $(k-1)$-dimensional subspace of C^*.*

Proof. If a $(k-1)$-dimensional subspace $S \subset C^*$ contains precisely m elements of $\mathcal{P}(C)$ then the weight of non-zero vectors belonging to the annihilator of S is equal to $n - m$. □

It follows from Lemma 2.23 that the minimal weight of C is not greater than $n - k + 1$. Also, the minimal weight is less than $n - k + 1$ if the elements of $\mathcal{P}(C)$ are not mutually distinct.

Corollary 2.5. *The following two conditions are equivalent:*

- $\mathcal{P}(C)$ *is a k-independent subset of $\mathcal{G}_1(C^*)$,*
- *the minimal weight of C is equal to $n - k + 1$.*

Lemma 2.22 and Corollary 2.5 give the following characterization of k-independent subsets of $\mathcal{G}_1(W)$ in terms of the minimal weight of linear $[n, k]_q$ codes.

Corollary 2.6. *Suppose that $\mathcal{P}$ is a subset of $\mathcal{G}_1(W)$ consisting of n elements and spanning the projective space or the projective line associated to W. Then $\mathcal{P}$ is a k-independent subset if and only if the minimal weight of the corresponding equivalence class of linear $[n, k]_q$ codes is equal to $n - k + 1$.*

Now we reformulate Proposition 2.4 in terms of linear codes. A linear $[n, k]_q$ code $C \subset V$ is said to be *linearly symmetric* if for every permutation δ on $\{1, \ldots, n\}$ there is a linear automorphism of C which can be extended to a linear monomial automorphism of V transferring each e_i to a scalar multiple of $e_{\delta(i)}$. Every code linearly equivalent to a linearly symmetric code is linearly symmetric.

Example 2.12. Recall that every linear $[n, 1]_q$ code is formed by all vectors of type

$$(a_1 c, \ldots, a_n c),$$

where $a_1, \ldots, a_n$ are non-zero elements of F. This is a linearly symmetric code.

Example 2.13. Let $x_1, \ldots, x_{n-1}$ be a base of the vector space F^{n-1} and $n \geq 3$. Consider the $(n-1)$-simplex consisting of

$$\langle x_1 \rangle, \ldots, \langle x_{n-1} \rangle \quad \text{and} \quad \langle x_1 + \cdots + x_{n-1} \rangle$$

The associated equivalence class of linear $[n, n-1]_q$ codes contains the code C formed by all vectors of type

$$(c_1, \ldots, c_{n-1}, c_1 + \cdots + c_{n-1}).$$

The projective system $\mathcal{P}(C)$ consists of $P_1, \ldots, P_n$, where P_i is the element of $\mathcal{G}_1(C^*)$ containing $e_i^*|_C$. Lemma 2.22 shows that $\mathcal{P}(C)$ is an $(n-1)$-simplex of $\mathcal{G}_1(C^*)$. Thus $\mathcal{P}(C)$ is a PGL-subset and for every permutation δ on $\{1, \ldots, n\}$ there is a linear automorphism s_δ of C^* transferring every P_i to $P_{\delta(i)}$. The contragradient $\check{s}_\delta \in \mathrm{GL}(C)$ can be extended to a linear monomial automorphism of V which sends every e_i to a scalar multiple of $e_{\delta(i)}$ (see the proof of the implication $(2) \implies (1)$ in Lemma 2.20). Therefore, C is linearly symmetric.

Example 2.14. Suppose that the characteristic of F is equal to 3, i.e. $q = 3^r$. Let x_1, x_2 be a base of F^2. Consider the harmonic set consisting of

$$\langle x_1 \rangle, \ \langle x_2 \rangle, \ \langle x_1 + x_2 \rangle, \ \langle x_1 - x_2 \rangle.$$

It defines the equivalence class of linear $[4, 2]_q$ codes containing the code formed by all vectors of type

$$(c_1, c_2, c_1 + c_2, c_1 - c_2).$$

By Lemma 2.22, the associated projective system is a harmonic set. This is a PGL-subset and, as in the previous example, we get a linearly symmetric code.

Lemma 2.24. *Let C be a linear $[n, k]_q$ code. If C is linearly symmetric then all elements of the projective system $\mathcal{P}(C)$ are coincident, i.e. $k = 1$, or they are distinct and form a PGL-subset of $\mathcal{G}_1(C^*)$.*

Proof. Denote by P_i the element of $\mathcal{P}(C)$ containing $e_i^* |_C$. Since C is linearly symmetric, for every permutation δ on $\{1, \dots, n\}$ there is a linear monomial automorphism u_δ of V transferring every e_i to a scalar multiple of $e_{\delta(i)}$ and such that C is invariant for u_δ. The contragradient of $v_\delta = u_\delta |_C$ is a linear automorphism of C^* sending every P_i to $P_{\delta(i)}$ (see the proof of the implication (1) $\implies$ (2) in Lemma 2.20). Thus $P_1, \dots, P_n$ form a PGL-subset if they are mutually distinct.

Suppose that $P_i = P_j$ for some distinct i, j. The case when $n = 2$ is trivial. If $n \geq 3$ then for every $k \neq i, j$ there is a permutation δ on $\{1, \dots, n\}$ such that $\delta(i) = k$ and $\delta(j) = j$. Then $\check{v}_\delta(P_i) = P_k$ and $\check{v}_\delta(P_j) = P_j$ which implies that P_k coincides with $P_i = P_j$. $\square$

Proposition 2.4 together with Lemma 2.24 give the following.

Corollary 2.7. *Let C be a linear $[n, k]_q$ code. If C is linearly symmetric then one of the following possibilities is realized:*

- *C is a linear $[n, 1]_q$ code,*
- *C is linearly equivalent to the linear $[n, n-1]_q$ code formed by all vectors of type*

$$(c_1, \dots, c_{n-1}, c_1 + \dots + c_{n-1}),$$

- *the characteristic of F is equal to 3 and C is linearly equivalent to the linear $[4, 2]_q$ code consisting of all vectors of type*

$$(c_1, c_2, c_1 + c_2, c_1 - c_2).$$

Problem 2.5. A linear $[n, k]_q$ code C is said to be *semilinearly symmetric* if for every permutation δ on $\{1, \ldots, n\}$ there is a semilinear automorphism of C which can be extended to a semilinear monomial automorphism of V transferring each e_i to a scalar multiple of $e_{\delta(i)}$. The description of semilinearly symmetric codes is equivalent to the description of PΓL-subsets in the projective spaces over finite fields, see Problem 2.4.

Chapter 3

Isometric embeddings of Grassmann graphs

The Grassmann graph $\Gamma_k(V)$ is formed by all k-dimensional subspaces of an n-dimensional vector space V over a division ring. The case when $k = 1, n - 1$ is trivial: any two distinct vertices of such Grassmann graph are adjacent. For this reason we will always suppose that $1 < k < n - 1$. It follows from well-known Chow's theorem [Chow (1949)] that every automorphism of $\Gamma_k(V)$ is induced by a semilinear automorphism of V or a semilinear isomorphism of V to V^* and the second possibility can be realized only in the case when $n = 2k$.

In this chapter we investigate embeddings of the Grassmann graph $\Gamma_k(V)$ in the Grassmann graph $\Gamma_{k'}(V')$, where V and V' are n-dimensional and n'-dimensional vector spaces over division rings. The main result states that every isometric embedding of $\Gamma_k(V)$ in $\Gamma_{k'}(V')$ is induced by a semilinear embedding of special type (see Section 1.5). It must be pointed out that the semilinear embeddings associated to isometric embeddings of Grassmann graphs are not necessarily strong. Using non-strong semilinear embeddings, we also construct non-isometric embeddings of Grassmann graphs. The description of all non-isometric embeddings is an open problem.

The above mentioned Chow's theorem is a simple consequence of our description of isometric embeddings of Grassmann graphs. There are many results closely connected to Chow's theorem. Some of them will be presented in Section 3.9 without proofs. In Chapter 6 we reformulate Chow's theorem in terms of semilinear isomorphisms of exterior powers, as an application, we describe automorphisms of Grassmann codes.

At the end of the chapter we prove the following remarkable result [Huang (1998)]: every bijective transformation of a Grassmannian sending adjacent vertices of the associated Grassmann graph to adjacent vertices of the Grassmann graph is an automorphism of this graph. The statement is trivial for Grassmannians of vector spaces over finite fields, but it is hard to prove in the general case. Huang's theorem is related to Kreuzer's example and Problem 1.1.

3.1   Graph theory

In this short section we recall some definitions from graph theory which will be used throughout the second part of the book.

A *graph* $\Gamma = (X, \sim)$ consists of a vertex set X (possibly infinite) together with a symmetric relation $\sim$ called *adjacency*. A pair of vertices $v, w \in X$ form an *edge* of the graph if $v \sim w$. In what follows we will always suppose that $v \not\sim v$ for every $v \in X$, i.e. our graph does not contain loops.

For every subset $X' \subset X$ the graph $\Gamma' = (X', \sim)$ is said to be the *restriction* of the graph Γ to the subset X'.

A *clique* in Γ is a subset of X where any two distinct vertices are adjacent (a subset consisting of one vertex is a clique). Every clique is contained in a maximal clique. This statement is trivial for finite graphs and it follows from Zorn's lemma in the infinite case.

Let $v, w \in X$. A *path* between v and w is a sequence of vertices

$$v = v_0, v_1, \dots, v_k = w \tag{3.1}$$

such that $v_{i-1} \sim v_i$ for every $i \in \{1, \dots, k\}$. The number k (the number of edges in the path) is called the *length* of this path. We define the *distance* $d(v, w)$ between v and w as the minimal length of a path connecting them. We say that (3.1) is a *geodesic* if $d(v, w) = k$.

A graph is *connected* if any two distinct vertices can be connected by a path. In such a graph the distance between any pair of vertices is defined. The *diameter* of a connected graph is the greatest distance between two vertices.

Two graphs $\Gamma = (X, \sim)$ and $\Gamma' = (X', \sim')$ are *isomorphic* if there is a bijection of X to X' preserving the adjacency relation in both directions, i.e. two vertices of Γ are adjacent if and only if the corresponding vertices of Γ' are adjacent. Such bijections are called *isomorphisms*. Isomorphisms of a graph to itself are said to be *automorphisms*.

An *embedding* of $\Gamma = (X, \sim)$ in $\Gamma' = (X', \sim')$ is an injection of X to X' such that adjacent vertices go to adjacent vertices and non-adjacent vertices go to non-adjacent vertices. Every surjective embedding is an isomorphism. An embedding is called *isometric* if it preserves the distance between vertices. Every embedding preserves distances 1 and 2. If the diameter of a graph is not greater than 2 then every embedding of this graph is isometric. We refer [Deza and Laurent (1997), Part III] for the general theory of graph isometric embeddings.

An embedding f of a graph Γ in a graph Γ' is said to be *rigid* if for every automorphism g of Γ there is an automorphism g' of Γ' such that

$$fg = g'f.$$

This is connected to the notion of GL-mapping discussed in Section 1.8.

3.2 Elementary properties of Grassmann graphs

Let V be a left vector space over a division ring. Suppose that $\dim V = n$ is finite. Let $k \in \{1, \ldots, n-1\}$. The *Grassmann graph* $\Gamma_k(V)$ is the graph whose vertex set is the Grassmannian $\mathcal{G}_k(V)$ and $S, U \in \mathcal{G}_k(V)$ are adjacent vertices if

$$\dim(S \cap U) = k - 1,$$

or equivalently,

$$\dim(S + U) = k + 1.$$

In the case when $k = 1, n - 1$, any two distinct vertices of $\Gamma_k(V)$ are adjacent.

Proposition 3.1. *The annihilator mapping induces an isomorphism between* $\Gamma_k(V)$ *and* $\Gamma_{n-k}(V^*)$.

Proof. Let $S, U \in \mathcal{G}_k(V)$. The equality

$$(S \cap U)^0 = S^0 + U^0$$

implies that

$$\dim(S \cap U) = k - 1 \iff \dim(S^0 + U^0) = n - k + 1.$$

In other words, S and U are adjacent vertices of $\Gamma_k(V)$ if and only if S^0 and U^0 are adjacent vertices of $\Gamma_{n-k}(V^*)$. $\qquad\square$

Proposition 3.2. *The graph* $\Gamma_k(V)$ *is connected and for any* $S, U \in \mathcal{G}_k(V)$ *the distance* $d(S, U)$ *is equal to*

$$k - \dim(S \cap U) = \dim(S + U) - k.$$

The diameter of $\Gamma_k(V)$ *is equal to* $\min\{k, n - k\}$.

Proof. Let $S, U \in \mathcal{G}_k(V)$. Then

$$k - \dim(S \cap U) = \dim(S + U) - k.$$

Suppose that this number is equal to m. Lemma 1.1 implies the existence of linearly independent vectors $x_1, \ldots, x_{k+m}$ such that

$$S = \langle x_1, \ldots, x_k \rangle \quad \text{and} \quad U = \langle x_{m+1}, \ldots, x_{k+m} \rangle.$$

Then

$$\langle x_{1+i}, \ldots, x_{k+i} \rangle, \quad i \in \{0, \ldots, m\}$$

form a path connecting S and U. Thus the graph $\Gamma_k(V)$ is connected and $d(S, U) \leq m$. Suppose that

$$S = S_0, S_1, \ldots, S_i = U$$

is a geodesic. Using induction, we show that

$$\dim(S_0 + \cdots + S_j) \leq k + j$$

for every $j \in \{1, \ldots, i\}$; in particular,

$$\dim(S_0 + \cdots + S_i) \leq k + i$$

which implies that the dimension of $S + U$ is not greater than $k + i$. Then

$$m = \dim(S + U) - k \leq i = d(S, U).$$

So, $d(S, U) = m$. The statement concerning the diameter follows directly from the distance formula.   □

Two vertices of $\Gamma_k(V)$ are said to be *opposite* if the distance between them is equal to the diameter.

Remark 3.1. It is not difficult to prove that for every geodesic $S_0, S_1, \ldots, S_m$ in $\Gamma_k(V)$ the inclusions

$$S_0 \cap S_m \subset S_i \subset S_0 + S_m$$

hold for every i. Also, the geodesic can be extended to a maximal geodesic connecting S_0 with a certain opposite vertex.

Recall that for any pair of subspaces $S, U \subset V$ such that

$$S \subset U \quad \text{and} \quad \dim S < k < \dim U$$

we denote by $[S, U]_k$ the set of all $P \in \mathcal{G}_k(V)$ satisfying

$$S \subset P \subset U.$$

We will write $[S\rangle_k$ or $\langle U]_k$ if $U = V$ or $S = 0$, respectively.

Subsets of type

$$[S, U]_k, \quad S \in \mathcal{G}_{k-1}(V), \ U \in \mathcal{G}_{k+1}(V)$$

are called *lines*. These are lines of the projective space Π_V or lines of the dual projective space Π_V^* if $k = 1$ or $k = n-1$, respectively. Two elements of $\mathcal{G}_k(V)$ are adjacent vertices of $\Gamma_k(V)$ if and only if there is a line containing them. Such a line is unique.

It was noted above that any two distinct vertices of $\Gamma_k(V)$ are adjacent if $k = 1, n-1$. Now we suppose that $1 < k < n-1$ and describe all maximal cliques of $\Gamma_k(V)$.

Example 3.1. Every subset of type

$$[S\rangle_k, \ S \in \mathcal{G}_{k-1}(V)$$

is called a *star*. This is a clique of $\Gamma_k(V)$. The star $[S\rangle_k$ (together with all lines contained in it) is a projective space isomorphic to $\Pi_{V/S}$.

Example 3.2. Every subset of type

$$\langle U]_k, \ U \in \mathcal{G}_{k+1}(V)$$

is said to be a *top*. This is a clique of $\Gamma_k(V)$. The top $\langle U]_k$ (together with all lines contained in it) is a projective space isomorphic to Π_U^*.

Proposition 3.3. *If $1 < k < n - 1$ then every maximal clique of $\Gamma_k(V)$ is a star or a top.*

Proof. It is sufficient to show that every clique of $\Gamma_k(V)$ is contained in a star or a top. If a clique consists of two elements $S, U \in \mathcal{G}_k(V)$ then it is contained in the star $[S \cap U\rangle_k$ and in the top $\langle S + U]_k$.

Let $\mathcal{X}$ be a clique of $\Gamma_k(V)$ containing more than two elements and let S_1, S_2 be distinct elements of $\mathcal{X}$. Suppose that $\mathcal{X}$ is not contained in a star. This implies the existence of $S_3 \in \mathcal{X}$ which does not contain $S_1 \cap S_2$, i.e.

$$\dim(S_1 \cap S_2 \cap S_3) \le k - 2. \tag{3.2}$$

We show that S_3 is contained in $N := S_1 + S_2$.

Suppose that $S_3 \not\subset N$. Then

$$\dim(S_3 \cap N) \le k - 1.$$

Since S_3 is adjacent to S_1 and S_2, the subspace $S_3 \cap N$ is $(k-1)$-dimensional and it is contained in both S_1 and S_2 which contradicts (3.2). So, $S_3 \subset N$.

Let $S \in \mathcal{G}_k(V) \setminus \langle N \rangle_k$. Then

$$\dim(S \cap N) \leq k - 1.$$

If S is adjacent to S_1, S_2, S_3 then $S \cap N$ is a $(k-1)$-dimensional subspace contained in each S_i. This is impossible by (3.2). Thus S is not adjacent to a certain S_i. This means that $S \notin \mathcal{X}$. Therefore, $\mathcal{X}$ is contained in the top $\langle N \rangle_k$. □

Lemma 3.1. *The annihilator mapping of $\mathcal{G}_k(V)$ to $\mathcal{G}_{n-k}(V^*)$ transfers stars to tops and tops to stars.*

Proof. Easy verification. □

The intersection of two distinct maximal cliques of $\Gamma_k(V)$ is empty or a one-element set or a line. The third possibility is realized only in the case when the cliques are of different types, i.e. one of them is a star and the other is a top, and the associated $(k-1)$-dimensional and $(k+1)$-dimensional subspaces are incident. The intersection of two distinct stars is non-empty if and only if the associated $(k-1)$-dimensional subspaces are adjacent vertices of $\Gamma_{k-1}(V)$. Similarly, the intersection of two distinct tops is non-empty if and only if the associated $(k+1)$-dimensional subspaces are adjacent vertices of $\Gamma_{k+1}(V)$.

As above, we suppose that $1 < k < n - 1$. Denote by $\mathrm{Cl}_k(V)$ the graph whose vertex set is

$$\mathcal{G}_{k-1}(V) \cup \mathcal{G}_{k+1}(V)$$

and whose edges are pairs

$$S \subset \mathcal{G}_{k-1}(V), \ U \in \mathcal{G}_{k+1}(V) \ \text{such that} \ S \subset U.$$

The vertices of $\mathrm{Cl}_k(V)$ can be identified with the maximal cliques of $\Gamma_k(V)$. Two maximal cliques correspond to adjacent vertices if and only if their intersection is a line.

Lemma 3.2. *The graph $\mathrm{Cl}_k(V)$ is connected. If $C_0, \ldots, C_i$ is a path in $\mathrm{Cl}_k(V)$ then C_0 and C_i belong to the same Grassmannian only in the case when i is even.*

Proof. Let S and U be distinct vertices of $\mathrm{Cl}_k(V)$. In the case when S and U both belong to $\mathcal{G}_{k-1}(V)$, we take any geodesic

$$S = S_0, S_1, \ldots, S_i = U$$

in $\Gamma_{k-1}(V)$. Since $S_{j-1} + S_j$ belongs to $\mathcal{G}_k(V)$ for every $j \in \{1, \ldots, i\}$, there exist

$$U_1, \ldots, U_i \in \mathcal{G}_{k+1}(V)$$

such that each U_j contains $S_{j-1} + S_j$. Then

$$S_0, U_1, S_1, \ldots, U_i, S_i$$

is a path in $\mathrm{Cl}_k(V)$ connecting S and U. The case when $S, U \in \mathcal{G}_{k+1}(V)$ is similar.

Now we suppose that $S \in \mathcal{G}_{k+1}(V)$ and $U \in \mathcal{G}_{k-1}(V)$. We take any $S' \in \mathcal{G}_{k-1}(V)$ contained in S. Then S and S' are adjacent vertices of $\mathrm{Cl}_k(V)$ and it was established above that there is a path in $\mathrm{Cl}_k(V)$ connecting S' with U.

So, $\mathrm{Cl}_k(V)$ is connected. The second statement of the lemma is obvious.

$\square$

If V is a vector space over a finite field then $\Gamma_k(V)$ is one of the classical examples of so-called *distance-regular* graphs [Brouwer, Cohen and Neumaier (1989), Section 9.3]. There is an interesting problem concerning the characterizing $\Gamma_k(V)$ by the parameters of distance-regular graphs. For many values of k this is possible [Metsch (1995)]. If $n = 2k + 1$ then there is a distance-regular graph with the same parameters as $\Gamma_k(V)$ and non-isomorphic to it [van Dam and Koolen (2005)]. For some cases the problem is still open.

Example 3.3 ([van Dam and Koolen (2005)]). Let $n = 2k + 1$. Let also H be an $(n - 1)$-dimensional subspace of V. The associated *twisted Grassmann graph* $\tilde{\Gamma}_k(V, H)$ has two types of vertices. The vertices of first type are elements of $\mathcal{G}_{k+1}(V)$ which are not contained in H. The vertices of second type are elements of $\mathcal{G}_{k-1}(H)$. Two vertices of the same type are adjacent if they are adjacent vertices of the corresponding Grassmann graph $\Gamma_{k+1}(V)$ or $\Gamma_{k-1}(H)$. Two vertices of different types are adjacent if they are incident subspaces of V. This graph is not isomorphic to $\Gamma_k(V)$. If V is a vector space over a finite field then $\Gamma_k(V)$ and $\tilde{\Gamma}_k(V, H)$ are distance regular graphs of the same parameters. The graph $\tilde{\Gamma}_k(V, H)$ has precisely the following four types of maximal cliques:

(1) Let $S \in \mathcal{G}_k(V)$. The dimension of $H \cap S$ is equal to $k - 1$ or k if $S \not\subset H$ or $S \subset H$, respectively. Denote by $\mathcal{C}(S)$ the set formed by all vertices of first type containing S and all vertices of second type contained in

$H \cap S$. This is a maximal clique of $\tilde{\Gamma}_k(V, H)$. If $S \not\subset H$ then $H \cap S$ is an element of $\mathcal{G}_{k-1}(H)$ and

$$\mathcal{C}(S) = [S\rangle_{k+1} \cup \{H \cap S\}.$$

In the case when $S \subset H$, we have

$$\mathcal{C}(S) = ([S\rangle_{k+1} \setminus [S, H]_{k+1}) \cup \langle S]_{k-1}.$$

(2) For every $S \in \mathcal{G}_{k-2}(H)$ the star $[S\rangle_{k-1}$ is a maximal clique of $\tilde{\Gamma}_k(V, H)$.
(3) If $U \in \mathcal{G}_{k+2}(V)$ is not contained in H then $H \cap U$ is $(k+1)$-dimensional and

$$\langle U]_{k+1} \setminus \{H \cap U\}$$

is a maximal clique of $\tilde{\Gamma}_k(V, H)$.
(4) If $U \in \mathcal{G}_{k+2}(V)$ is not contained in H and S is a $(k-1)$-dimensional subspace of $H \cap U$ then

$$([S, U]_{k+1} \setminus \{H \cap U\}) \cup \{S\}$$

is a maximal clique of $\tilde{\Gamma}_k(V, H)$.

See [van Dam and Koolen (2005); Fujisaki, Koolen and Tagami (2006)] for more information.

3.3 Embeddings

From this moment we suppose that V and V' are left vector spaces over division rings R and R' (respectively) and the dimensions $\dim V = n$, $\dim V' = n'$ both are finite. We will consider embeddings of $\Gamma_k(V)$ in $\Gamma_{k'}(V')$. The following facts are obvious:

- if $k = 1, n - 1$ then every embedding of $\Gamma_k(V)$ in $\Gamma_{k'}(V')$ is a bijection to a clique of $\Gamma_{k'}(V')$,
- there are no embeddings of $\Gamma_k(V)$ in $\Gamma_{k'}(V')$ in the case when $1 < k < n - 1$ and $k' = 1, n' - 1$.

For these reasons we will suppose that

$$1 < k < n - 1 \quad \text{and} \quad 1 < k' < n' - 1$$

which means that n and n' both are not less than 4.

If $m \geq k$ then every semilinear m-embedding $l : V \to V'$ induces the mapping

$$(l)_k : \mathcal{G}_k(V) \to \mathcal{G}_k(V')$$

transferring each $S \in \mathcal{G}_k(V)$ to $\langle l(S) \rangle$. Some properties of such mappings were established in Section 1.5. Now we prove the following.

Proposition 3.4. *If $l : V \to V'$ is a semilinear m-embedding and $k \leq m-2$ then $(l)_k$ is an embedding of $\Gamma_k(V)$ in $\Gamma_k(V')$.*

Proof. By Proposition 1.7, the mapping $(l)_k$ is injective. Let $S, U \in \mathcal{G}_k(V)$. If S and U are adjacent vertices of $\Gamma_k(V)$ then

$$\dim(S + U) = k + 1 < m$$

and the dimension of

$$\langle l(S + U) \rangle = \langle l(S) \rangle + \langle l(U) \rangle \tag{3.3}$$

is equal to $k + 1$ which means that $\langle l(S) \rangle$ and $\langle l(U) \rangle$ are adjacent vertices of $\Gamma_k(V')$. Suppose that S and U are non-adjacent vertices of $\Gamma_k(V)$. In this case, we have

$$\dim(S + U) \geq k + 2$$

and there is a $(k + 2)$-dimensional subspace $T \subset S + U$. Since $k + 2 \leq m$, $\langle l(T) \rangle$ is a $(k + 2)$-dimensional subspace contained in (3.3). Hence the dimension of (3.3) is not less than $k + 2$ which guarantees that $\langle l(S) \rangle$ and $\langle l(U) \rangle$ are non-adjacent vertices of $\Gamma_k(V')$. $\qquad \square$

Every embedding of $\Gamma_k(V)$ in $\Gamma_{k'}(V')$ transfers maximal cliques of $\Gamma_k(V)$ to not necessarily maximal cliques of $\Gamma_{k'}(V')$, i.e. subsets in maximal cliques.

Lemma 3.3. *For every embedding of $\Gamma_k(V)$ in $\Gamma_{k'}(V')$ the following assertions are fulfilled:*

(1) *Each maximal clique of $\Gamma_{k'}(V')$ contains at most one image of a maximal clique of $\Gamma_k(V)$.*

(2) *The image of every maximal clique of $\Gamma_k(V)$ is contained in precisely one maximal clique of $\Gamma_{k'}(V')$.*

Proof. Let f be an embedding of $\Gamma_k(V)$ in $\Gamma_{k'}(V')$.

(1). Let $\mathcal{X}$ and $\mathcal{Y}$ be distinct maximal cliques of $\Gamma_k(V)$. We choose $X \in \mathcal{X}$ and $Y \in \mathcal{Y}$ which are not adjacent. Then $f(X)$ and $f(Y)$ are non-adjacent vertices of $\Gamma_{k'}(V')$ which means that there is no maximal clique of $\Gamma_{k'}(V')$ containing both $f(\mathcal{X})$ and $f(\mathcal{Y})$.

(2). Suppose that $\mathcal{X}$ is a maximal clique of $\Gamma_k(V)$ and $f(\mathcal{X})$ is contained in two distinct maximal cliques of $\Gamma_{k'}(V')$. The intersection of these cliques is a line (since $f(\mathcal{X})$ contains more than one element and the intersection of two distinct maximal cliques of a Grassmann graph is empty or a one-element set or a line). So, there exist $S \in \mathcal{G}_{k'-1}(V')$ and $U \in \mathcal{G}_{k'+1}(V')$ such that

$$f(\mathcal{X}) \subset [S, U]_{k'}. \tag{3.4}$$

We take any maximal clique $\mathcal{Y}$ of $\Gamma_k(V)$ intersecting $\mathcal{X}$ precisely in a line and consider a maximal clique $\mathcal{Y}'$ of $\Gamma_{k'}(V')$ containing $f(\mathcal{Y})$. The intersection of $f(\mathcal{X})$ and $\mathcal{Y}'$ contains more than one element. Then (3.4) guarantees that the intersection of $[S, U]_{k'}$ and $\mathcal{Y}'$ contains more than one element. This is possible only in the case when $\mathcal{Y}'$ is the star $[S\rangle_{k'}$ or the top $\langle U]_{k'}$. In each of these cases, the maximal clique $\mathcal{Y}'$ contains $f(\mathcal{X})$ and $f(\mathcal{Y})$ which contradicts (1). □

In the previous section we introduced the graph $\mathrm{Cl}_k(V)$ whose vertices are identified with maximal cliques of $\Gamma_k(V)$ and two vertices are adjacent if the intersection of the corresponding maximal cliques is a line.

Lemma 3.4. *Every embedding of $\Gamma_k(V)$ in $\Gamma_{k'}(V')$ induces an injection of the vertex set of $\mathrm{Cl}_k(V)$ to the vertex set of $\mathrm{Cl}_{k'}(V')$ which transfers adjacent vertices to adjacent vertices.*

Proof. Let f be an embedding of $\Gamma_k(V)$ in $\Gamma_{k'}(V')$. It follows from the second part of Lemma 3.3 that f induces a mapping of the vertex set of $\mathrm{Cl}_k(V)$ to the vertex set of $\mathrm{Cl}_{k'}(V')$. The first part of Lemma 3.3 guarantees that this mapping is injective. If two vertices of $\mathrm{Cl}_k(V)$ are adjacent then the corresponding maximal cliques $\mathcal{X}$ and $\mathcal{Y}$ are intersecting in a line. The latter implies that $f(\mathcal{X}) \cap f(\mathcal{Y})$ contains more than one element. Hence the intersection of the maximal cliques of $\Gamma_{k'}(V')$ containing $f(\mathcal{X})$ and $f(\mathcal{Y})$ is a line and the corresponding vertices of $\mathrm{Cl}_{k'}(V')$ are adjacent. □

By Lemma 3.4, the mapping between the vertex sets of $\mathrm{Cl}_k(V)$ and $\mathrm{Cl}_{k'}(V')$ induced by an embedding of $\Gamma_k(V)$ in $\Gamma_{k'}(V')$ transfers every path of $\mathrm{Cl}_k(V)$ to a path of $\mathrm{Cl}_{k'}(V')$. Using Lemma 3.2 we show that for this mapping one the following possibilities is realized:

- the image of $\mathcal{G}_{k-1}(V)$ is contained in $\mathcal{G}_{k'-1}(V')$ and the image of $\mathcal{G}_{k+1}(V)$ is contained in $\mathcal{G}_{k'+1}(V')$,
- the image of $\mathcal{G}_{k-1}(V)$ is contained in $\mathcal{G}_{k'+1}(V')$ and the image of $\mathcal{G}_{k+1}(V)$ is contained in $\mathcal{G}_{k'-1}(V')$.

So, we get the following.

Proposition 3.5. *For every embedding of $\Gamma_k(V)$ in $\Gamma_{k'}(V')$ one of the following possibilities is realized:*

(A) *stars go to subsets of stars and tops go to subsets of tops,*
(B) *stars go to subsets of tops and tops go to subsets of stars.*

We say that an embedding is of *type* (A) or of *type* (B) if the corresponding possibility is realized.

Example 3.4. The embedding of $\Gamma_k(V)$ in $\Gamma_k(V')$ described in Proposition 3.4 is of type (A).

3.4 Isometric embeddings

In this section we start to investigate isometric embeddings of $\Gamma_k(V)$ in $\Gamma_{k'}(V')$. As in the previous section, we suppose that

$$1 < k < n - 1 \quad \text{and} \quad 1 < k' < n' - 1.$$

The existence of isometric embeddings implies that

$$\min\{k, n - k\} \leq \min\{k', n' - k'\}, \tag{3.5}$$

i.e. the diameter of $\Gamma_k(V)$ is not greater than the diameter of $\Gamma_{k'}(V')$.

Proposition 3.6. *If $l : V \to V'$ is a semilinear m-embedding and $m \geq 2k$ then $(l)_k$ is an isometric embedding of $\Gamma_k(V)$ in $\Gamma_k(V')$.*

Proof. By Proposition 3.4, the mapping $(l)_k$ is an embedding of $\Gamma_k(V)$ in $\Gamma_k(V')$. For any $S, U \in \mathcal{G}_k(V)$ we have

$$\dim(S + U) \leq 2k \leq m.$$

Then the subspaces

$$S + U \quad \text{and} \quad \langle l(S + U) \rangle = \langle l(S) \rangle + \langle l(U) \rangle$$

are of the same dimension which implies that $d(S, U) = d(\langle l(S) \rangle, \langle l(U) \rangle)$.
$\square$

Let f be a mapping of $\mathcal{G}_k(V)$ to $\mathcal{G}_{k'}(V')$. We distinguish the following three mappings related to f:

- $f^* : \mathcal{G}_k(V) \to \mathcal{G}_{n'-k'}(V'^*)$ sending every P to $f(P)^0$,

- $f_* : \mathcal{G}_{n-k}(V^*) \to \mathcal{G}_{k'}(V')$ sending every P to $f(P^0)$
- $\check{f} : \mathcal{G}_{n-k}(V^*) \to \mathcal{G}_{n'-k'}(V'^*)$ sending every P to $f(P^0)^0$.

If $n = n'$ and f is induced by a strong semilinear embedding $u : V \to V'$ then the mapping $\check{f}$ is induced by the contragradient $\check{u} : V^* \to V'^*$, see Proposition 1.12.

Lemma 3.5. *If f is an isometric embedding of $\Gamma_k(V)$ in $\Gamma_{k'}(V')$ then f^*, f_* and $\check{f}$ are isometric embeddings of $\Gamma_k(V)$ in $\Gamma_{n'-k'}(V'^*)$, $\Gamma_{n-k}(V^*)$ in $\Gamma_{k'}(V')$ and $\Gamma_{n-k}(V^*)$ in $\Gamma_{n'-k'}(V'^*)$, respectively. The embeddings f and $\check{f}$ are of the same type, i.e. (A) or (B). The embeddings f and f^* are of different types and the same holds for the embeddings f and f_*.*

Proof. This follows from the fact that the annihilator mapping induces isomorphisms between Grassmann graphs transferring stars to tops and tops to stars. □

Example 3.5. Let $s : V \to V'^*$ be a semilinear m-embedding such that $m \geq 2k$. The mapping

$$(s)_k^* : \mathcal{G}_k(V) \to \mathcal{G}_{n'-k}(V')$$

$$S \to s(S)^0$$

is the composition of $(s)_k$ and the annihilator mapping of $\mathcal{G}_k(V'^*)$ to $\mathcal{G}_{n'-k}(V')$. By Proposition 3.6 and Lemma 3.5, this is an isometric embedding of $\Gamma_k(V)$ in $\Gamma_{n'-k}(V')$. This embedding is of type (B).

If $k = 2, n - 2$ then the diameter of $\Gamma_k(V)$ is equal to 2 and every embedding of $\Gamma_k(V)$ is isometric. The following example shows that there are isometric embeddings induced by non-strong semilinear embeddings and non-isometric embeddings of Grassmann graphs.

Example 3.6. Let l be a semilinear n-embedding of an $(n+m)$-dimensional vector space W in an n-dimensional vector space W', see Example 1.18. By Proposition 3.4, the mapping $(l)_k$ is an embedding of $\Gamma_k(W)$ in $\Gamma_k(W')$ for every $k \leq n - 2$. Proposition 3.6 implies that this embedding is isometric if $k \leq n - k$. In the case when

$$n - k < k \leq n - 2,$$

the diameters of $\Gamma_k(W)$ and $\Gamma_k(W')$ are equal to

$$\min\{k, n + m - k\} \text{ and } n - k,$$

respectively. The inequality

$$\min\{k, n + m - k\} > n - k$$

shows that the embedding $(l)_k$ is non-isometric.

Suppose that $S \in \mathcal{G}_s(V)$ is contained in $U \in \mathcal{G}_u(V)$ and $u - s \geq 2$. Denote by Φ_S^U the mapping of $\mathcal{G}(U/S)$ to $\mathcal{G}(V)$ transferring every subspace of U/S to the corresponding subspace of V. In the case when $U = V$ or $S = 0$, this mapping will be denoted by Φ_S or Φ^U, respectively. It is clear that Φ^U is the identity injection of $\mathcal{G}(U)$ to $\mathcal{G}(V)$. An easy verification shows that for every integer $i \in \{1, \ldots, u - s - 1\}$ the restriction of Φ_S^U to $\mathcal{G}_i(U/S)$ is an isometric embedding of $\Gamma_i(U/S)$ in $\Gamma_{s+i}(V)$. The image of this embedding is $[S, U]_{s+i}$.

In the following two examples we suppose that $k \leq n - k$. The condition (3.5) implies that

$$k \leq \min\{k', n' - k'\}. \tag{3.6}$$

Example 3.7. Let $S \in \mathcal{G}_{k'-k}(V')$. By (3.6), we have

$$\dim(V'/S) = n' - k' + k \geq 2k. \tag{3.7}$$

Suppose that $l : V \to V'/S$ is a semilinear m-embedding and $m \geq 2k$ (note that the inequality (3.7) is necessary to the existence of such a semilinear embedding). The mapping $(l)_k$ is an isometric embedding of $\Gamma_k(V)$ in $\Gamma_k(V'/S)$. The restriction of Φ_S to $\mathcal{G}_k(V'/S)$ is an isometric embedding of $\Gamma_k(V'/S)$ in $\Gamma_{k'}(V')$. The composition $\Phi_S(l)_k$ is an isometric embedding of $\Gamma_k(V)$ in $\Gamma_{k'}(V')$. This embedding is of type (A).

Example 3.8. The condition (3.6) implies that $k' + k \leq n'$. Suppose that $U \in \mathcal{G}_{k'+k}(V')$ and $s : V \to U^*$ is a semilinear m-embedding such that $m \geq 2k$. By Example 3.5, $(s)_k^*$ is an isometric embedding of $\Gamma_k(V)$ in $\Gamma_{k'}(U)$. Then $\Phi^U(s)_k^*$ is an isometric embedding of $\Gamma_k(V)$ in $\Gamma_{k'}(V')$. This embedding is of type (B).

Example 3.9. Suppose that $S \in \mathcal{G}_{k'-k}(V')$ is contained in $U \in \mathcal{G}_{k'+k}(V')$ and $n = 2k$. Then

$$\dim(U/S) = 2k = n,$$

i.e. the vector spaces V and U/S are of the same dimension. For every strong semilinear embedding $l : V \to U/S$ the mapping $\Phi_S^U(l)_k$ is an isometric embedding of $\Gamma_k(V)$ in $\Gamma_{k'}(V')$. If $s : V \to (U/S)^*$ is a strong semilinear embedding then $(s)_k^*$ is an isometric embedding of $\Gamma_k(V)$ in $\Gamma_k(U/S)$ and $\Phi_S^U(s)_k^*$ is an isometric embedding of $\Gamma_k(V)$ in $\Gamma_{k'}(V')$.

Theorem 3.1 ([Pankov 2 (2012)]). *Let f be an isometric embedding of $\Gamma_k(V)$ in $\Gamma_{k'}(V')$. If $k \le n - k$ then one of the following possibilities is realized:*

- *f is of type (A) then there is $S \in \mathcal{G}_{k'-k}(V')$ such that*

$$f = \Phi_S(l)_k,$$

     where $l : V \to V'/S$ is a semilinear m-embedding and $m \ge 2k$;
- *f is of type (B) then there is $U \in \mathcal{G}_{k'+k}(V')$ such that*

$$f = \Phi^U(s)_k^*,$$

     where $s : V \to U^*$ is a semilinear m-embedding and $m \ge 2k$.

In the case when $n = 2k \ge 4$, there exist incident

$$S \in \mathcal{G}_{k'-k}(V') \quad and \quad U \in \mathcal{G}_{k'+k}(V')$$

such that f is induced by a strong semilinear embedding of V in U/S (type (A)) or a strong semilinear embedding of V in $(U/S)^*$ (type (B)), see Example 3.9.

Remark 3.2. For the case when $k = k'$ the above result was obtained in [Kosiorek, Matras and Pankov (2008)].

For every semilinear isomorphism $l : V \to V'$ the mapping $(l)_k$ is an isomorphism of $\Gamma_k(V)$ to $\Gamma_k(V')$. If s is a semilinear isomorphism of V to V'^* then $(s)_k^*$ is an isomorphism of $\Gamma_k(V)$ to $\Gamma_{n-k}(V')$. In the case when $n = 2k$, this is an isomorphism between $\Gamma_k(V)$ and $\Gamma_k(V')$.

As a simple consequence of Theorem 3.1, we get classical Chow's theorem [Chow (1949)] concerning isomorphisms of Grassmann graphs.

Corollary 3.1. *If $\Gamma_k(V)$ is isomorphic to $\Gamma_{k'}(V')$ then*

$$n = n' \quad and \quad k' = k \quad or \quad k' = n - k.$$

Every isomorphism of $\Gamma_k(V)$ to $\Gamma_k(V')$ is induced by a semilinear isomorphism of V to V' or a semilinear isomorphism of V to V'^* and the second possibility can be realized only in the case when $n = 2k$. If $n \ne 2k$ then every isomorphism of $\Gamma_k(V)$ to $\Gamma_{n-k}(V')$ is induced by a semilinear isomorphism of V to V'^*.

Proof. Let f be an isomorphism of $\Gamma_k(V)$ to $\Gamma_{k'}(V')$. Suppose that $k \le n - k$. By Theorem 3.1, one of the following possibilities is realized:

(1) $f = \Phi_S(l)_k$, where $S \in \mathcal{G}_{k'-k}(V')$ and $l : V \to V'/S$ is a semilinear m-embedding with $m \geq 2k$;

(2) $f = \Phi^U(s)_k^*$, where $U \in \mathcal{G}_{k+k'}(V')$ and $s : V \to U^*$ is a semilinear m-embedding with $m \geq 2k$.

Since f is bijective, we have $S = 0$ and $f = (l)_k$ in the first case. Proposition 1.8 implies that l is a semilinear isomorphism of V to V'. Then $n = n'$ and $k = k'$.

Similarly, we get that $U = V'$ and $f = (s)_k^*$ in the second case. By Proposition 1.8, s is a semilinear isomorphism V to V'^*. Then $n = n'$ and $k' = n - k$.

Consider the case when $n - k < k$. The mapping $\check{f}$ is an isomorphism of $\Gamma_{n-k}(V^*)$ to $\Gamma_{n'-k'}(V'^*)$. Since $n - k < n - (n - k)$, we apply the above arguments to $\check{f}$ and establish that $n = n'$ and k' is equal to k or $n - k$. If $k = k'$ then $\check{f}$ is induced by a semilinear isomorphism $u : V^* \to V'^*$ and Proposition 1.11 guarantees that f is induced by the contragradient $\check{u}$. If $k' = n - k$ then f^* is an isomorphism of $\Gamma_k(V)$ to $\Gamma_k(V'^*)$, it is induced by a semilinear isomorphism $s : V \to V'^*$ and we have $f = (s)_k^*$. $\qquad\square$

In Theorem 3.1, we require that $k \leq n - k$. Now we give some remarks concerning the general case.

Example 3.10. Suppose that
$$k \leq k' \quad \text{and} \quad n - k \leq n' - k'.$$
If $S \in \mathcal{G}_{k'-k}(V')$ then
$$\dim(V'/S) = n' - k' + k \geq n.$$
For every strong semilinear embedding $l : V \to V'/S$ the mapping $\Phi_S(l)_k$ is an isometric embedding of $\Gamma_k(V)$ in $\Gamma_{k'}(V')$. There exists $U \in \mathcal{G}_{n+k'-k}(V')$ such that l can be consider as a strong semilinear embedding of V in U/S and we can write $\Phi_S^U(l)_k$ for our isometric embedding.

Example 3.11. Suppose that
$$n \leq k + k' \leq n'.$$
If $U \in \mathcal{G}_{k'+k}(V')$ and $s : V \to U^*$ is a strong semilinear embedding then $\Phi^U(s)_k^*$ is an isometric embedding of $\Gamma_k(V)$ in $\Gamma_{k'}(V')$. Let S be the annihilator of $s(V)$ in U. Then
$$S \in \mathcal{G}_{k'+k-n}(V') \quad \text{and} \quad \dim(U/S) = n.$$
Since every element of $s(V)$ is a linear functional of U whose kernel contains S, the embedding s induces a strong semilinear embedding $u : V \to (U/S)^*$. An easy verification shows that $\Phi_S^U(u)_k^*$ coincides with $\Phi^U(s)_k^*$.

Remark 3.3. Let f be an isometric embedding of $\Gamma_k(V)$ in $\Gamma_{k'}(V')$. Suppose that $n - k < k$. By Lemma 3.5, f_* is an isometric embedding of $\Gamma_{n-k}(V^*)$ in $\Gamma_{k'}(V')$. Since $n - k < n - (n - k)$, this is one of the isometric embeddings described in Theorem 3.1. The condition (3.5) implies that

$$n - k \leq \min\{k', n' - k'\}.$$

There are the following two possibilities:

(1) f is of type (A) then f_* is of type (B) and there exists $U \in \mathcal{G}_{k'+n-k}(V')$ such that

$$f_* = \Phi^U(s)^*_{n-k},$$

where $s : V^* \to U^*$ is a semilinear m-embedding and $m \geq 2(n - k)$. We have

$$f = \Phi^U(s)^*_{n-k}\mathrm{A},$$

where A is the annihilator mapping of $\mathcal{G}_k(V)$ to $\mathcal{G}_{n-k}(V^*)$. Suppose that s is a strong semilinear embedding. By Example 3.11, s induces a strong semilinear embedding $u : V^* \to (U/S)^*$, where $S \in \mathcal{G}_{k'-k}(V')$ is the annihilator of $s(V)$ in U, and

$$f = \Phi^U_S(u)^*_{n-k}\mathrm{A}.$$

Since $\dim(U/S) = n$, Proposition 1.12 implies that

$$f = \Phi^U_S(\check{u})_k,$$

where $\check{u} : V \to U/S$ is the contragradient of u. Clearly, we can consider $\check{u}$ as a strong semilinear embedding of V in V'/S and write $f = \Phi_S(\check{u})_k$.

(2) f is of type (B) then f_* is of type (A) and there exists $S \in \mathcal{G}_{k'+k-n}(V')$ such that

$$f_* = \Phi_S(v)_{n-k},$$

where $v : V^* \to V'/S$ is a semilinear m-embedding and $m \geq 2(n - k)$. We have

$$f = \Phi_S(v)_{n-k}\mathrm{A},$$

as above, A is the annihilator mapping of $\mathcal{G}_k(V)$ to $\mathcal{G}_{n-k}(V^*)$. Suppose that v is a strong semilinear embedding. Then it can be considered as a strong semilinear embedding of V^* in U/S, where $U \in \mathcal{G}_{k+k'}(V')$. As in the previous case, $\dim(U/S) = n$. By Proposition 1.12, we have

$$(v)_{n-k}\mathrm{A} = \mathrm{A}'(\check{v})_k = (\check{v})^*_k,$$

where $\breve{v} : V \to (U/S)^*$ is the contragradient of $v : V^* \to U/S$ and A$'$ is the annihilator mapping of $\mathcal{G}_k((U/S)^*)$ to $\mathcal{G}_{n-k}(U/S)$. Thus

$$f = \Phi_S^U(\breve{v})_k^*.$$

Since every linear functional of U/S can be considered as a linear functional of U whose kernel contains S, the strong semilinear embedding $\breve{v} : V \to (U/S)^*$ induces a strong semilinear embedding $w : V \to U^*$. It is clear that $(w)_k^*$ coincides with $\Phi_S(\breve{v})_k^*$ and $f = \Phi^U(w)_k^*$.

Therefore, if f_* is induced by a strong semilinear embedding of V^* in a vector space then f is induced by a strong semilinear embedding of V in a vector space. In the general case, f cannot be reduced to a semilinear embedding of V in a certain vector space.

In some special cases, all isometric embeddings of Grassmann graphs are induced by semilinear isomorphisms.

Corollary 3.2. *Suppose that $R = R'$ is a division ring satisfying the following condition:*

(*) *every endomorphism of the division ring is an automorphism and every homomorphism to the opposite division rings is an isomorphism.*

If f is an isometric embedding of $\Gamma_k(V)$ in $\Gamma_{k'}(V')$ then one of the following possibilities is realized:

- *f is of type* (A) *then $0 \le k' - k \le n' - n$ and there exist incident subspaces*

$$S \in \mathcal{G}_{k'-k}(V'), \quad U \in \mathcal{G}_{k'-k+n}(V')$$

such that $f = \Phi_S^U(l)_k$, where $l : V \to U/S$ is a semilinear isomorphism;
- *f is of type* (B) *then $n \le k' + k \le n'$ and there exist incident subspaces*

$$S \in \mathcal{G}_{k'+k-n}(V'), \quad U \in \mathcal{G}_{k'+k}(V')$$

such that $f = \Phi_S^U(s)_k^*$, where $s : V \to (U/S)^*$ is a semilinear isomorphism.

Proof. First we consider the case when $k \le n - k$.

If f is an embedding of type (A) then it is induced by a semilinear embedding $l : V \to V'/S$, where $S \in \mathcal{G}_{k'-k}(V')$. It follows from (*) that l is a semilinear isomorphism to an n-dimensional subspace of V'/S and we get the claim.

If f is an embedding of type (B) then it is induced by a semilinear embedding $\tilde{s} : V \to U^*$, where $U \in \mathcal{G}_{k'+k}(V')$. By (*), $\tilde{s}$ is a semilinear isomorphism to the n-dimensional subspace $\tilde{s}(V) \subset U^*$. Let $S \in \mathcal{G}_{k'+k-n}(V')$ be the annihilator of $\tilde{s}(V)$ in U. Then $\tilde{s}$ induces a semilinear isomorphism $s : V \to (U/S)^*$ which is as required.

In the case when $k > n - k$, we apply the above arguments to f_* and use Remark 3.3. $\qquad\Box$

Example 3.12. The condition (*) holds for the fields $\mathbb{Q}$ and $\mathbb{R}$, the field of p-adic numbers and all finite fields.

Example 3.13. Now we suppose that $R = \mathrm{GF}(p^m)$ and $R' = \mathrm{GF}(q^l)$, where p and q are prime numbers. By Theorem 3.1 and Example 1.7, the existence of isometric embeddings of $\Gamma_k(V)$ in $\Gamma_{k'}(V')$ implies that $p = q$ and m divides l. The case when $p = q$ and $m = l$, i.e. $R = R'$, is covered by Corollary 3.2. Suppose that $p = q$ and $l = dm$ with $d > 1$. By Examples 1.18 and 3.6, there exist isometric embeddings of $\Gamma_k(V)$ in $\Gamma_{k'}(V')$ induced by non-strong semilinear embeddings if d is sufficiently large.

3.5   Proof of Theorem 3.1

Let f be an isometric embedding of $\Gamma_k(V)$ in $\Gamma_{k'}(V')$ and $k \le n - k$. Then

$$k \le \min\{k', n' - k'\},$$

i.e. the diameter of $\Gamma_k(V)$ is not greater than the diameter of $\Gamma_{k'}(V')$.

Case (A). Suppose that f is an embedding of type (A). Then f transfers stars to subsets of stars. By Section 3.3, there is an injective mapping

$$f_{k-1} : \mathcal{G}_{k-1}(V) \to \mathcal{G}_{k'-1}(V')$$

such that

$$f([P\rangle_k) \subset [f_{k-1}(P)\rangle_{k'} \quad \forall\, P \in \mathcal{G}_{k-1}(V),$$

i.e. f sends the star $[P\rangle_k$ to a subset in the star $[f_{k-1}(P)\rangle_{k'}$. If $P \in \mathcal{G}_{k-1}(V)$ and $Q \in \mathcal{G}_k(V)$ then

$$P \in \langle Q]_{k-1} \iff Q \in [P\rangle_k \implies f(Q) \in [f_{k-1}(P)\rangle_{k'}$$

$$\iff f_{k-1}(P) \in \langle f(Q)]_{k'-1}$$

and we have

$$f_{k-1}(\langle Q]_{k-1}) \subset \langle f(Q)]_{k'-1} \quad \forall\, Q \in \mathcal{G}_k(V).$$

Thus f_{k-1} transfers tops to subsets of tops. This means that f_{k-1} maps adjacent vertices of $\Gamma_{k-1}(V)$ to adjacent vertices of $\Gamma_{k'-1}(V')$ (since f_{k-1} is injective and for any two adjacent vertices of $\Gamma_{k-1}(V)$ there is a top containing them).

Lemma 3.6. *The mapping f_{k-1} is an isometric embedding of $\Gamma_{k-1}(V)$ in $\Gamma_{k'-1}(V')$. This embedding is of type* (A).

Proof. Since f_{k-1} is injective and transfers adjacent vertices to adjacent vertices, every path of $\Gamma_{k-1}(V)$ goes to a path of $\Gamma_{k'-1}(V')$ and we have

$$d(P,Q) \geq d(f_{k-1}(P), f_{k-1}(Q))$$

for any $P, Q \in \mathcal{G}_{k-1}(V)$. We need to prove the inverse inequality.

The condition $2k \leq n$ implies the existence of $P', Q' \in \mathcal{G}_k(V)$ such that

$$P \subset P', \quad Q \subset Q' \quad \text{and} \quad P \cap Q = P' \cap Q'.$$

Then

$$d(P,Q) = k - 1 - \dim(P \cap Q) = k - 1 - \dim(P' \cap Q') = d(P',Q') - 1$$

and we get

$$d(P,Q) = d(P',Q') - 1. \tag{3.8}$$

Since f_{k-1} is induced by f, we have

$$f_{k-1}(P) \subset f(P') \quad \text{and} \quad f_{k-1}(Q) \subset f(Q')$$

which guarantees that

$$\dim(f_{k-1}(P) \cap f_{k-1}(Q)) \leq \dim(f(P') \cap f(Q')). \tag{3.9}$$

Using (3.8) and (3.9), we obtain the required inequality

$$d(P,Q) = d(P',Q') - 1 = d(f(P'), f(Q')) - 1 = k' - 1 - \dim(f(P') \cap f(Q'))$$

$$\leq k' - 1 - \dim(f_{k-1}(P) \cap f_{k-1}(Q)) = d(f_{k-1}(P), f_{k-1}(Q)).$$

So, f_{k-1} is an isometric embedding of $\Gamma_{k-1}(V)$ in $\Gamma_{k'-1}(V')$. Since f_{k-1} sends tops to subsets of tops, this is an embedding of type (A). $\qquad\square$

Step by step, we construct a sequence of isometric embeddings f_i of $\Gamma_i(V)$ in $\Gamma_{k'-k+i}(V')$ with $i = k, \ldots, 1$ such that $f_k = f$ and

$$f_i([P\rangle_i) \subset [f_{i-1}(P)\rangle_{k'-k+i} \quad \forall P \in \mathcal{G}_{i-1}(V)$$

if $i > 1$. The latter inclusion implies that

$$f_{i-1}(\langle Q]_{i-1}) \subset \langle f_i(Q)]_{k'-k+i-1} \quad \forall Q \in \mathcal{G}_i(V); \tag{3.10}$$

in particular, we have

$$f_1(\langle Q]_1) \subset \langle f_2(Q)]_{k'-k+1} \quad \forall\, Q \in \mathcal{G}_2(V). \tag{3.11}$$

Lemma 3.7. *If $Q \in \mathcal{G}_i(V)$ and $i \in \{2,\ldots,k\}$ then*

$$f_1(\langle Q]_1) \subset \langle f_i(Q)]_{k'-k+1}$$

and there is no proper subspace of $f_i(Q)$ which contains all elements of $f_1(\langle Q]_1)$.

Proof. We prove the statement by induction. In the case when $i = 2$, the required inclusion is (3.11). If all elements of $f_1(\langle Q]_1)$ are contained in a proper subspace of $f_2(Q) \in \mathcal{G}_{k'-k+2}(V')$ then they are coincident and f_1 is non-injective.

Let $i \geq 3$. Then $\langle Q]_1$ is the union of all $\langle P]_1$ such that $P \in \langle Q]_{i-1}$. By inductive hypothesis,

$$f_1(\langle Q]_1) = \bigcup_{P \in \langle Q]_{i-1}} f_1(\langle P]_1)$$

is contained in

$$\bigcup_{P \in \langle Q]_{i-1}} \langle f_{i-1}(P)]_{k'-k+1}.$$

It follows from (3.10) that

$$f_{i-1}(P) \subset f_i(Q) \ \text{ if } \ P \in \langle Q]_{i-1}$$

and we get the required inclusion.

Suppose that all elements of $f_1(\langle Q]_1)$ are contained in a proper subspace Q' of $f_i(Q)$. We have

$$\dim Q' \leq k' - k + i - 1.$$

Since

$$\dim f_{i-1}(P) = k' - k + i - 1$$

for all $P \in \mathcal{G}_{i-1}(V)$ and f_{i-1} is injective, there is $P \in \langle Q]_{i-1}$ such that $Q' \cap f_{i-1}(P)$ is a proper subspace of $f_{i-1}(P)$. Then all elements of $f_1(\langle P]_1)$ are contained in $Q' \cap f_{i-1}(P)$ which contradicts the inductive hypothesis. $\square$

Since f_1 is an isometric embedding of $\Gamma_1(V)$ in $\Gamma_{k'-k+1}(V')$, the image of f_1 is a clique in $\Gamma_{k'-k+1}(V')$. This clique is not contained in a top. Indeed, if there is $U \in \mathcal{G}_{k'-k+2}(V')$ such that the image of f_1 is a subset in the top $\langle U]_{k'-k+1}$ then Lemma 3.7 implies that $f_2(Q) = U$ for every $Q \in \mathcal{G}_2(V)$ and f_2 is non-injective. Therefore, there exists $S \in \mathcal{G}_{k'-k}(V')$ such that the image of f_1 is contained in the star $[S\rangle_{k'-k+1}$.

The inclusion (3.11) shows that f_1 transfers every line of Π_V to a subset of a line contained in $[S\rangle_{k'-k+1}$. The star $[S\rangle_{k'-k+1}$ (together with all lines contained in it) can be naturally identified with the projective space $\Pi_{V'/S}$ and there is a mapping

$$f' : \mathcal{G}_1(V) \to \mathcal{G}_1(V'/S)$$

such that

$$f_1 = \Phi_S f'.$$

It is clear that f' is injective and transfers all lines of Π_V to subsets contained in lines of $\Pi_{V'/S}$. The image of f' is not contained in a line of $\Pi_{V'/S}$ (since the image of f_1 is not contained in a top). By Corollary 2.1, f' is induced by a semilinear 2-embedding $l : V \to V'/S$. Then

$$f_1 = \Phi_S(l)_1 \quad \text{and} \quad f_1(\langle Q]_1) \subset \langle \Phi_S(\langle l(Q)\rangle)]_{k'-k+1}$$

for every subspace $Q \subset V$. If $Q \in \mathcal{G}_k(V)$ then, by Lemma 3.7, we have

$$f_1(\langle Q]_1) \subset \langle f(Q)]_{k'-k+1}.$$

Since $f(Q)$ is k'-dimensional and the dimension of $\Phi_S(\langle l(Q)\rangle)$ is not greater than k', the second part of Lemma 3.7, implies that

$$f(Q) = \Phi_S(\langle l(Q)\rangle) \quad \forall\, Q \in \mathcal{G}_k(V). \tag{3.12}$$

This equality guarantees that l is a semilinear k-embedding.

Show that l is a semilinear $(2k)$-embedding. Every independent $(2k)$-element subset $X \subset V$ can be presented as the disjoint union of two independent k-element subsets X_1 and X_2. Then $\langle X_1 \rangle$ and $\langle X_2 \rangle$ are opposite vertices of $\Gamma_k(V)$ and

$$d(f(\langle X_1 \rangle), f(\langle X_2 \rangle)) = d(\langle X_1 \rangle, \langle X_2 \rangle) = k$$

which implies that

$$f(\langle X_1 \rangle) \cap f(\langle X_2 \rangle) = S.$$

Since

$$f(\langle X_i \rangle) = \Phi_S(\langle l(X_i) \rangle), \quad i \in \{1, 2\},$$

$\langle l(X_1)\rangle$ and $\langle l(X_2)\rangle$ are k-dimensional subspaces of V'/S whose intersection is 0. This means that the dimension of

$$\langle l(X)\rangle = \langle l(X_1 \cup X_2)\rangle = \langle l(X_1)\rangle + \langle l(X_2)\rangle$$

is equal to $2k$. Hence $l(X)$ is an independent subset.

So, l is a semilinear m-embedding of V in V'/S with $m \geq 2k$ and, by (3.12), we have

$$f = \Phi_S(l)_k.$$

If $n = 2k$ then l is a strong semilinear embedding and the image of l is contained in U/S, where $U \in \mathcal{G}_{k'+k}(V')$. In this case, we consider l as a strong semilinear embedding of V in U/S and get $f = \Phi_S^U(l)_k$.

Case (B). Suppose that f is an embedding of type (B). It follows from Lemma 3.5 that f^* is an isometric embedding of $\Gamma_k(V)$ in $\Gamma_{n'-k'}(V'^*)$ of type (A) and, by the arguments given above, its image is contained in

$$[S']_{n'-k'}, \quad S' \in \mathcal{G}_{n'-k'-k}(V'^*).$$

This means that the image of f is a subset of $\langle U]_{k'}$, where $U \in \mathcal{G}_{k'+k}(V')$ is the annihilator of S'. Thus f is an isometric embedding of $\Gamma_k(V)$ in $\Gamma_{k'}(U)$. Consider the mapping which sends every $P \in \mathcal{G}_k(V)$ to the annihilator of $f(P)$ in U. This is an isometric embedding of $\Gamma_k(V)$ in $\Gamma_k(U^*)$ of type (A). It is induced by a semilinear m-embedding $s : V \to U^*$ with $m \geq 2k$. Then

$$f = \Phi^U(s)_k^*.$$

In the case when $n = 2k$, the image of f^* is contained in

$$[S', U']_{n'-k'}, \quad S' \in \mathcal{G}_{n'-k'-k}(V'^*), \; U' \in \mathcal{G}_{n'-k'+k}(V'^*).$$

Then the image of f is a subset of $[S, U]_{k'}$, where

$$S \in \mathcal{G}_{k'-k}(V') \quad \text{and} \quad U \in \mathcal{G}_{k'+k}(V')$$

are the annihilators of U' and S', respectively. It is clear that

$$f = \Phi_S^U f',$$

where f' is an isometric embedding of $\Gamma_k(V)$ in $\Gamma_k(U/S)$ of type (B). We have

$$\dim(U/S) = 2k = n$$

and f'^* is an isometric embedding of $\Gamma_k(V)$ in $\Gamma_k((U/S)^*)$ of type (A). Then f'^* is induced by a strong semilinear embedding

$$s : V \to (U/S)^*$$

which means that $f' = (s)_k^*$ and $f = \Phi_S^U(s)_k^*$.

3.6 Equivalence of isometric embeddings

There is the following natural equivalence relation on the set of all isometric embeddings of $\Gamma_k(V)$ in $\Gamma_{k'}(V')$: two embeddings f and g are equivalent if there exist an automorphism h of $\Gamma_k(V)$ and an automorphism h' of $\Gamma_{k'}(V')$ such that

$$g = h'fh. \tag{3.13}$$

By Chow's theorem (Corollary 3.1), every automorphism of $\Gamma_k(V)$ is induced by a semilinear automorphism of V or a semilinear isomorphism of V to V^* and the second possibility can be realized only in the case when $n = 2k$. For this reason, we will say that two isometric embeddings f and g of $\Gamma_k(V)$ in $\Gamma_{k'}(V')$ are *equivalent* if there exist an automorphism h of $\Gamma_k(V)$ and an automorphism h' of $\Gamma_{k'}(V')$ induced by semilinear automorphisms of V and V' (respectively) and satisfying (3.13). If $n \neq 2k$ and $n' \neq 2k'$ then this equivalence relation coincides with the equivalence relation considered above.

Lemma 3.8. *The following four conditions are equivalent:*

- f *and* g *are equivalent isometric embeddings of* $\Gamma_k(V)$ *in* $\Gamma_{k'}(V')$,
- f_* *and* g_* *are equivalent isometric embeddings of* $\Gamma_{n-k}(V^*)$ *in* $\Gamma_{k'}(V')$,
- f^* *and* g^* *are equivalent isometric embeddings of* $\Gamma_k(V)$ *in* $\Gamma_{n'-k'}(V'^*)$,
- $\check{f}$ *and* $\check{g}$ *are equivalent isometric embeddings of* $\Gamma_{n-k}(V^*)$ *in* $\Gamma_{n'-k'}(V'^*)$.

Proof. Let u and u' be semilinear automorphisms of V and V', respectively. Then the following four conditions are equivalent:

- $g = (u')_{k'} f(u)_k$,
- $g_* = (u')_{k'} f_*(\check{u})_{n-k}$,
- $g^* = (\check{u}')_{n'-k'} f^*(u)_k$,
- $\check{g} = (\check{u}')_{n'-k'} \check{f}(\check{u})_{n-k}$.

This gives the claim. □

Lemma 3.9. *Let* l *and* s *be semilinear* m-*embeddings of* V *in* V' *such that* $m \geq 2k$. *Then the isometric embeddings* $(l)_k$ *and* $(s)_k$ *are equivalent if and only if* l *and* s *are equivalent.*

Proof. If there exist $u \in \Gamma L(V)$ and $u' \in \Gamma L(V')$ such that $s = u'lu$ then

$$(s)_k = (u')_k(l)_k(u)_k.$$

Conversely, the latter equality implies that $(s)_k = (u'lu)_k$ and, by Proposition 1.9, we have $s = u''lu$, where u'' is a scalar multiple of u'. $\qquad\square$

Let f be an isometric embedding of $\Gamma_k(V)$ in $\Gamma_{k'}(V')$ and $k \le n - k$. If f is an embedding of type (A) then it is induced by a certain semilinear embedding $l : V \to V'/S$, where $S \in \mathcal{G}_{k'-k}(V')$. Let σ be the associated homomorphism of R to R'. Any other semilinear embedding of V in V'/S inducing f is a scalar multiple of l and the associated homomorphism $\sigma' : R \to R'$ coincides with $\delta\sigma$, where δ is an inner automorphism of R'. The homomorphisms σ and σ' are equivalent and we denote by $\mathfrak{H}(f)$ the corresponding equivalence class of homomorphisms. If f is an embedding of type (B) then f^* is an isometric embedding of $\Gamma_k(V)$ in $\Gamma_{n'-k'}(V'^*)$ of type (A) and we define

$$\mathfrak{H}(f) := \mathfrak{H}(f^*),$$

i.e. $\mathfrak{H}(f)$ is an equivalence class of homomorphisms of R to R'^*.

Lemma 3.9 shows that the classification of isometric embeddings of $\Gamma_k(V)$ in $\Gamma_k(V')$ can be reduced to the classification of the associated semilinear embeddings. If $n = 2k$ then isometric embeddings are induced by strong semilinear embeddings and the required classification easily follows from Proposition 1.4.

Proposition 3.7. *Let f and g be isometric embeddings of $\Gamma_k(V)$ in $\Gamma_{k'}(V')$ of the same type and $n = 2k$. Then f and g are equivalent if and only if $\mathfrak{H}(f) = \mathfrak{H}(g)$.*

Proof. If our embeddings are of type (A) then there exist

$$S, \tilde{S} \in \mathcal{G}_{k'-k}(V') \quad \text{and} \quad U, \tilde{U} \in \mathcal{G}_{k'+k}(V')$$

such that $S \subset U$, $\tilde{S} \subset \tilde{U}$ and

$$f = \Phi_S^U(l)_k, \quad g = \Phi_{\tilde{S}}^{\tilde{U}}(s)_k,$$

where $l : V \to U/S$ and $s : V \to \tilde{U}/\tilde{S}$ are strong semilinear embeddings.

Suppose that for some $u \in \Gamma L(V)$ and $u' \in \Gamma L(V')$ we have

$$g = (u')_{k'} f(u)_k. \tag{3.14}$$

Then u' transfers S and U to $\tilde{S}$ and $\tilde{U}$, respectively. Hence u' induces a semilinear isomorphism $w : U/S \to \tilde{U}/\tilde{S}$ (Lemma 1.3). An easy verification shows that

$$(s)_k = (w)_k (l)_k (u)_k.$$

As in the proof of Lemma 3.9, we establish that $s = w'lu$, where w' is a scalar multiple of w. The latter implies that the homomorphisms of R to R' associated to l and s are equivalent, i.e. $\mathfrak{H}(f) = \mathfrak{H}(g)$.

Conversely, if $\mathfrak{H}(f) = \mathfrak{H}(g)$ then, as in the proof of Proposition 1.4, we construct $u \in \Gamma L(V)$ and a semilinear isomorphism $w : U/S \to \tilde{U}/\tilde{S}$ such that $s = wlu$. Let W and $\tilde{W}$ be complements of S and $\tilde{S}$ in U and $\tilde{U}$, respectively. By Remark 1.3, for every $x \in W$ there is unique $\tilde{x} \in \tilde{W}$ satisfying

$$w(x + S) = \tilde{x} + \tilde{S}$$

and the correspondence $x \to \tilde{x}$ is a semilinear isomorphism of W to $\tilde{W}$. We extend it to a semilinear automorphism $u' \in \Gamma L(V')$ sending S to $\tilde{S}$. After that we establish the equality (3.14).

If f and g are embeddings of type (B) then f^* and g^* are isometric embeddings of $\Gamma_k(V)$ in $\Gamma_{n'-k'}(V'^*)$ of type (A). By the arguments given above, f^* and g^* are equivalent if and only if $\mathfrak{H}(f^*) = \mathfrak{H}(f)$ coincides with $\mathfrak{H}(g^*) = \mathfrak{H}(g)$. The required statement follows from Lemma 3.8. $\quad\square$

Example 3.14. Suppose that W is an $(n+1)$-dimensional real vector space and W' is an n-dimensional vector space over the field of rational functions $\mathbb{R}(t)$. By Example 1.19, there are infinitely many equivalence classes of semilinear n-embeddings of W in W' over the identity homomorphism of $\mathbb{R}$ to $\mathbb{R}(t)$. Then Lemma 3.8 implies the existence of infinitely many equivalence classes of isometric embeddings of $\Gamma_k(W)$ in $\Gamma_k(W')$ corresponding to the same equivalence class of homomorphisms of $\mathbb{R}$ to $\mathbb{R}(t)$.

3.7 Linearly rigid isometric embeddings

Let f be a rigid isometric embedding of $\Gamma_k(V)$ in $\Gamma_k(V')$, i.e. for every automorphism g of $\Gamma_k(V)$ there is an automorphism g' of $\Gamma_k(V')$ satisfying

$$fg = g'f. \tag{3.15}$$

If $k \leq n - k$ and f is an embedding of type (A) then it is induced by a σ-linear embedding $l : V \to V'$. In the case when $n \neq 2k$ and $n' \neq 2k$, for every $u \in \Gamma L(V)$ there exists $u' \in \Gamma L(V')$ such that

$$(l)_k(u)_k = (u')_k(l)_k.$$

Then $(lu)_k = (u'l)_k$ and Proposition 1.9 implies that $lu = \tilde{u}l$, where $\tilde{u}$ is a scalar multiple of u'. Thus for every $\gamma \in \mathrm{Aut}(R)$ there exists $\gamma' \in \mathrm{Aut}(R')$

such that

$$\sigma\gamma = \gamma'\sigma$$

(the reader can check that the same holds in the case when f is an embedding of type (B)). By Remark 1.8, the latter is not true in general.

So, if R and R' form a pair of special type (for example, $R = \mathbb{C}$ and $R' = \mathbb{H}$ or R is one of the fields considered in Examples 1.2, 1.3 and $R' = \mathbb{R}$) then isometric embeddings of $\Gamma_k(V)$ in $\Gamma_k(V')$ exist, but they are not rigid. For this reason, we will consider the following version of rigidity.

We say that an isometric embedding f of $\Gamma_k(V)$ in $\Gamma_{k'}(V')$ is *linearly rigid* if for every automorphism g of $\Gamma_k(V)$ induced by a linear automorphism of V there is an automorphism g' of $\Gamma_{k'}(V')$ induced by a linear automorphism of V' and satisfying (3.15).

Lemma 3.10. *The following four conditions are equivalent:*

- *f is a linearly rigid isometric embedding of $\Gamma_k(V)$ in $\Gamma_{k'}(V')$,*
- *f_* is a linearly rigid isometric embedding of $\Gamma_{n-k}(V^*)$ in $\Gamma_{k'}(V')$,*
- *f^* io a linearly rigid isometric embedding of $\Gamma_k(V)$ in $\Gamma_{n'-k'}(V'^*)$,*
- *$\check{f}$ is a linearly rigid isometric embedding of $\Gamma_{n-k}(V^*)$ in $\Gamma_{n'-k'}(V'^*)$.*

Proof. Let u and u' be linear automorphisms of V and V', respectively. Then the following four conditions are equivalent:

- $f(u)_k = (u')_{k'} f$,
- $f_*(\check{u})_{n-k} = (u')_{k'} f_*$,
- $f^*(u)_k = (\check{u}')_{n'-k'} f^*$,
- $\check{f}(\check{u})_{n-k} = (\check{u}')_{n'-k'} \check{f}$.

Our statement follows from the fact that the contragradient mapping is an isomorphism of $\mathrm{GL}(V)$ to $\mathrm{GL}(V^*)$ and $\mathrm{GL}(V')$ to $\mathrm{GL}(V'^*)$. $\square$

Example 3.15. Recall that every strong semilinear embedding $l : V \to V'$ is a GL-mapping (Section 1.8). This implies that $(l)_k$ is a linearly rigid isometric embedding of $\Gamma_k(V)$ in $\Gamma_k(V')$ for every k. If $s : V \to V'^*$ is a strong semilinear embedding then, by Lemma 3.10, $(s)_k^*$ is a linearly rigid isometric embedding of $\Gamma_k(V)$ in $\Gamma_{n'-k}(V')$ for every k.

Example 3.16. Suppose that

$$S \in \mathcal{G}_{k'-k}(V') \quad \text{and} \quad l : V \to V'/S$$

is a strong semilinear embedding. Then $(l)_k$ is a linearly rigid isometric embedding of $\Gamma_k(V)$ in $\Gamma_k(V'/S)$, i.e. for every $u \in \mathrm{GL}(V)$ there exists $\tilde{u} \in \mathrm{GL}(V'/S)$ such that

$$(l)_k (u)_k = (\tilde{u})_k (l)_k.$$

We take any complement W to S. For every $x \in W$ there is unique $\tilde{x} \in W$ satisfying

$$\tilde{u}(x + S) = \tilde{x} + S$$

and the correspondence $x \to \tilde{x}$ is a linear automorphism of W (Remark 1.3). We extend it to a linear automorphism $u' \in \mathrm{GL}(V')$ preserving S. It is easy to see that

$$\Phi_S(l)_k (u)_k = (u')_{k'} \Phi_S(l)_k.$$

Therefore, $\Phi_S(l)_k$ is a linearly rigid isometric embedding of $\Gamma_k(V)$ in $\Gamma_{k'}(V')$. Similarly, if

$$U \in \mathcal{G}_{k+k'}(V') \quad \text{and} \quad s : V \to U^*$$

is a strong semilinear embedding then $(s)_k^*$ is a linearly rigid isometric embedding of $\Gamma_k(V)$ in $\Gamma_{k'}(U)$ which implies that $\Phi^U(s)_k^*$ is a linearly rigid isometric embedding of $\Gamma_k(V)$ in $\Gamma_{k'}(V')$.

Theorem 3.1 and the above examples show that every isometric embedding of $\Gamma_k(V)$ in $\Gamma_{k'}(V')$ is linearly rigid if $n = 2k$. We use Proposition 1.13 to prove the following.

Proposition 3.8. *If R' is a field then for every linearly rigid isometric embedding f of $\Gamma_k(V)$ in $\Gamma_{k'}(V')$ one of the following possibilities is realized:*

- *there exists $S \in \mathcal{G}_{k'-k}(V')$ such that*

$$f = \Phi_S(l)_k,$$

  where $l : V \to V'/S$ is a strong semilinear embedding;
- *there exists $U \in \mathcal{G}_{k'+k}(V')$ such that*

$$f = \Phi^U(s)_k^*,$$

  where $s : V \to U^*$ is a strong semilinear embedding.

Proof. The case when $n = 2k$ is obvious and we suppose that $k < n - k$.

If f is an embedding of type (A) then there exists $S \in \mathcal{G}_{k'-k}(V')$ such that $f = \Phi_S(l)_k$, where $l : V \to V'/S$ is a semilinear embedding. For every $u \in \mathrm{GL}(V)$ there is $u' \in \mathrm{GL}(V')$ satisfying

$$\Phi_S(l)_k (u)_k = (u')_{k'} \Phi_S(l)_k.$$

Then u' preserves S and induces a linear automorphism $\tilde{u}$ of V'/S. An easy verification shows that

$$(l)_k (u)_k = (\tilde{u})_k (l)_k.$$

Hence $(lu)_k = (\tilde{u}l)_k$ and, by Proposition 1.9, there exists a non-zero scalar $a \in R'$ such that

$$lu = a\tilde{u}l.$$

Since R' is a field, $a\tilde{u}$ is a linear automorphism of V'. So, l is a GL-mapping and Proposition 1.13 implies that l is a strong semilinear embedding.

If f is an embedding of type (B) then there exists $U \in \mathcal{G}_{k'+k}(V')$ such that $f = \Phi^U(s)_k^*$, where $s : V \to U^*$ is a semilinear embedding. For every $u \in \mathrm{GL}(V)$ there exists $u' \in \mathrm{GL}(V')$ such that

$$\Phi^U(s)_k^*(u)_k = (u')_{k'} \Phi^U(s)_k^*.$$

Then U is invariant for u' and we have

$$(s)_k^*(u)_k = (\tilde{u})_{k'}(s)_k^*,$$

where $\tilde{u}$ is the restriction of u' to U. Thus $(s)_k^*$ is a linearly rigid isometric embedding of $\Gamma_k(V)$ in $\Gamma_{k'}(U)$ and, by Lemma 3.10, $(s)_k$ is a linearly rigid isometric embedding of $\Gamma_k(V)$ in $\Gamma_k(U^*)$. As above, we establish that s is a strong semilinear embedding.

Consider the case when $k > n-k$. By Lemma 3.10, f_* is a linearly rigid isometric embedding of $\Gamma_{n-k}(V^*)$ in $\Gamma_{k'}(V')$. Since $n - k < n - (n - k)$, we can apply the above arguments to f_* and use Remark 3.3. $\qquad\square$

Problem 3.1. We cannot prove the same statement if R' is non-commutative. In this case, for a non-zero linear mapping $l : V \to V'$ the scalar multiple al is linear if and only if a belongs to the center of R'.

3.8   Remarks on non-isometric embeddings

If $k = 2, n - 2$ then the diameter of $\Gamma_k(V)$ is equal to 2 and every embedding of $\Gamma_k(V)$ in $\Gamma_{k'}(V')$ is isometric. By Example 3.6, non-isometric embeddings of Grassmann graphs exist.

Let f be an embedding of $\Gamma_k(V)$ in $\Gamma_{k'}(V')$ and

$$3 \le k \le n - k.$$

Suppose that f is an embedding of type (A). Then, by Section 3.3, it induces an injection

$$f_{k-1} : \mathcal{G}_{k-1}(V) \to \mathcal{G}_{k'-1}(V').$$

We have

$$f_{k-1}(\langle U]_{k-1}) \subset \langle f(U)]_{k'-1} \quad \forall\, U \in \mathcal{G}_k(V) \tag{3.16}$$

which means that f_{k-1} transfers tops to subsets of tops (if $k' \geq 3$) or to subsets contained in lines of $\Pi_{V'}$ (if $k' = 2$). This guarantees that f_{k-1} sends adjacent vertices of $\Gamma_{k-1}(V)$ to adjacent vertices of $\Gamma_{k'-1}(V')$.

So, if $k' = 2$ then f induces an injection of $\mathcal{G}_{k-1}(V)$ to $\mathcal{G}_1(V')$ transferring tops to subsets contained in lines of $\Pi_{V'}$. The description of such mappings is an open problem.

From this moment we suppose that $k' \geq 3$. Then f_{k-1} maps every maximal clique of $\Gamma_{k-1}(V)$ to a subset in a maximal clique of $\Gamma_{k'-1}(V')$.

Lemma 3.11. *The mapping f_{k-1} transfers stars to subsets of stars.*

Proof. Suppose that S is a star of $\mathcal{G}_{k-1}(V)$ whose image is contained in a top $\langle U']_{k'-1}$. We choose $U, Q \in \mathcal{G}_k(V)$ such that the tops $\langle U]_{k-1}$ and $\langle Q]_{k-1}$ intersect S in some lines. By (3.16), the tops $\langle f(U)]_{k'-1}$ and $\langle f(Q)]_{k'-1}$ intersect $\langle U']_{k'-1}$ in subsets containing more than one element. This is possible only in the case when

$$f(U) = U' = f(Q)$$

which contradicts the fact that f is injective. $\qquad\square$

Since the intersection of two distinct stars contains at most one element and f_{k-1} is injective, the image of a star is contained in a unique star. Therefore, f_{k-1} induces a mapping

$$f_{k-2} : \mathcal{G}_{k-2}(V) \to \mathcal{G}_{k'-2}(V')$$

such that

$$f_{k-2}(\langle U]_{k-2}) \subset \langle f_{k-1}(U)]_{k'-2} \quad \forall\, U \in \mathcal{G}_{k-1}(V). \tag{3.17}$$

This mapping is not necessarily injective, but (3.17) shows that the images of any two adjacent vertices of $\Gamma_{k-2}(V)$ are adjacent or coincident vertices of $\Gamma_{k'-2}(V')$.

Suppose that $k = 3$. Then $f_{k-1} = f_2$ and $f_{k-2} = f_1$ are mappings of $\mathcal{G}_2(V)$ and $\mathcal{G}_1(V)$ to $\mathcal{G}_{k'-1}(V')$ and $\mathcal{G}_{k'-2}(V')$, respectively. Since any two distinct vertices of $\Gamma_1(V)$ are adjacent, the image of f_1 is a clique of $\Gamma_{k'-2}(V)$.

Lemma 3.12. *If f_1 is non-constant then the image of f_1 cannot be contained in a top.*

Proof. Suppose that the image of f_1 is contained in a certain top $\langle T]_{k'-2}$. If $U \in \mathcal{G}_2(V)$ and $f_2(U) \neq T$ then (3.17) implies that
$$f_1(P) = T \cap f_2(U)$$
for every P belonging to the line $\langle U]_1$. Therefore, the restriction of f_1 to the line of Π_V defined by U is constant. Since f_2 is injective, there is at most one $U \in \mathcal{G}_2(V)$ satisfying $f_2(U) = T$. It is not difficult to prove that f_1 is constant. $\square$

Suppose that f_1 is non-constant. By Lemma 3.12, the image of f_1 is contained in a certain star $[S\rangle_{k'-2}$. Then
$$f = \Phi_S f',$$
where f' is a mapping of $\mathcal{G}_1(V)$ to $\mathcal{G}_1(V'/S)$. It follows from (3.17) that f' sends lines of Π_V to subsets in lines of $\Pi_{V'/S}$. Lemma 3.12 guarantees that the image of f' is not contained in a line of $\Pi_{V'/S}$. Since f_2 is injective and f_1 is non-constant, Lemma 2.2 implies that f' is injective. Thus f' satisfies the conditions of Corollary 2.1, i.e. f' is induced by a semilinear injection $l : V \to V'/S$. As in the proof of Theorem 3.1 (Section 2.5), we establish that l is a semilinear 5-embedding and $f = \Phi_S(l)_3$.

If f_1 is constant, i.e. there is $Q \in \mathcal{G}_{k'-2}(V')$ such that $f_1(P) = Q$ for every $P \in \mathcal{G}_1(V)$, then the image of f is contained in $[Q\rangle_{k'}$. Note that the restriction of $\Gamma_{k'}(V')$ to $[Q\rangle_{k'}$ is isomorphic to $\Gamma_2(V'/Q)$.

So, we get the following.

Proposition 3.9. *Let f be an embedding of $\Gamma_3(V)$ in $\Gamma_{k'}(V')$ and $n \geq 6$. If f is an embedding of type* (A) *then one of the following possibilities is realized:*

- *there exists $S \in \mathcal{G}_{k'-3}(V')$ such that*
$$f = \Phi_S(l)_3,$$
  where $l : V \to V'/S$ is a semilinear 5-embedding,
- *there exists $Q \in \mathcal{G}_{k'-2}(V')$ such that the image of f is contained in $[Q\rangle_{k'}$, i.e. f is an embedding of $\Gamma_3(V)$ in $\Gamma_2(V'/Q)$.*

Remark 3.4. There is a similar result for embeddings of type (B). We leave all details for the readers.

Problem 3.2. Are there embeddings of $\Gamma_3(V)$ in $\Gamma_2(V')$ (the second possibility of Proposition 3.9)?

The case when $k \geq 4$ is more complicated. By the arguments given above, the restriction of f_{k-2} to every star of $\mathcal{G}_{k-2}(V)$ is injective or constant, but there is no any information on the global behavior of f_{k-2}.

3.9 Some results related to Chow's theorem

Let V be a left vector space over a division ring. The dimension of V is denoted by n and assumed to be finite.

If $k = 1, n - 1$ then any two distinct vertices of $\Gamma_k(V)$ are adjacent and every bijective transformation of $\mathcal{G}_k(V)$ is an automorphism of $\Gamma_k(V)$. In the case when $1 < k < n - 1$, Chow's theorem [Chow (1949)] (Corollary 3.1) states that every automorphism of $\Gamma_k(V)$ is induced by a semilinear automorphism of V or a semilinear isomorphism of V to V^* and the second possibility can be realized only in the case when $n = 2k$. Thus the automorphism group of $\Gamma_k(V)$ coincides with $\mathrm{P\Gamma L}(V)$ if $n \neq 2k$ or there are not semilinear isomorphisms of V to V^*. Semilinear isomorphisms between V and V^* exist only in the case when the division ring associated to V is isomorphic to the opposite division ring (for example, if V is a vector space over a field or a vector space over the quaternion division ring $\mathbb{H}$). Suppose that the latter condition holds and $n = 2k$. Then $\mathrm{P\Gamma L}(V)$ is a normal subgroup in the automorphism group of $\Gamma_k(V)$ and the corresponding quotient group is isomorphic to $\mathbb{Z}_2$. In addition, if there exists a semilinear isomorphism $s : V \to V^*$ such that s^* is a scalar multiple of s (this holds for vector spaces over fields and quaternion vector spaces, see Example 1.20) then $(s)_k^*$ is an involution and the automorphism group of $\Gamma_k(V)$ can be presented as the semidirect product of $\mathrm{P\Gamma L}(V)$ and $\mathbb{Z}_2$.

Now we suppose that V is a vector space over a finite field. Let G be the group of all bijective transformations of the Grassmannian $\mathcal{G}_k(V)$. This group is finite and contains $\mathrm{P\Gamma L}(V)$ as a proper subgroup. In the case when $n \geq 3$, the following assertions are fulfilled [Ustimenko (1978)]:

- if $n \neq 2k$ then $\mathrm{P\Gamma L}(V)$ is a maximal subgroup of G,
- if $n = 2k$ then the subgroup of G generated by $\mathrm{P\Gamma L}(V)$ and the transformations induced by semilinear isomorphisms of V to V^* (the semidirect product of $\mathrm{P\Gamma L}(V)$ and $\mathbb{Z}_2$) is maximal.

For $n = 2$ the same fails (the Mathieu groups are counterexamples). Chow's theorem is a simple consequence of this result. Indeed, if $1 < k < n - 1$ then the automorphism group of $\Gamma_k(V)$ is a proper subgroup of G containing $\mathrm{P\Gamma L}(V)$; it also contains the transformations induced by semilinear isomorphisms of V to V^* if $n = 2k$. It is clear that the same arguments do not work for Grassmannians of vector spaces over infinite division rings.

Let us consider the graph whose vertex set consists of all $(m \times n)$-matrices over a division ring and two such matrices M_1 and M_2 are adjacent vertices of the graph if

$$\text{rank}(M_1 - M_2) = 1.$$

All automorphisms of this graph are described in [Hua (1951)]. This result is known as the Fundamental Theorem of Geometry of Rectangular Matrices. Chow's theorem can be deduced from Hua's theorem and conversely (see [Wan (1996), Section 3.9] for the references). Also, we refer [Šemrl (2014)] for the recent development of the topic. In this paper adjacency preserving mappings between $(m \times n)$-matrices and $(m' \times n')$-matrices are investigated.

By [Blunck and Havlicek (2005); Havlicek and Pankov (2005)], the adjacency relation of $\Gamma_k(V)$ can be characterized in terms of the opposite relation (as above, we suppose that $1 < k < n$). Recall that two vertices of $\Gamma_k(V)$ are opposite if the distance between them is equal to the graph diameter. For any distinct $S, U \in \mathcal{G}_k(V)$ the following two conditions are equivalent:

(1) S and U are adjacent vertices of $\Gamma_k(V)$,
(2) there exists $N \in \mathcal{G}_k(V) \backslash \{S, U\}$ such that every vertex of $\Gamma_k(V)$ opposite to N is opposite to S or opposite to U.

This means that every bijective transformation of $\mathcal{G}_k(V)$ preserving the relation to be opposite in both directions (two vertices of $\Gamma_k(V)$ are opposite if and only if their images are opposite) is an automorphism of $\Gamma_k(V)$.

Let m be an integer less than the diameter of $\Gamma_k(V)$, i.e.

$$m < \min\{k, n - k\}.$$

For every subset $\mathcal{X} \subset \mathcal{G}_k(V)$ we define

$$\mathcal{X}^m := \{S \in \mathcal{G}_k(V) : d(S, U) \le m \ \ \forall U \in \mathcal{X}\}.$$

For example, if S and U are adjacent vertices of $\Gamma_k(V)$ then

$$\{S, U\}^1 = \langle S + U \rangle_k \cup [S \cap U \rangle_k \ \text{ and } \ \{S, U\}^{11} = [S \cap U, S \mid U]_k.$$

The adjacency relation of $\Gamma_k(V)$ can be characterized in terms of the distance not greater than m [Lim (2010)]. For any distinct $S, U \in \mathcal{G}_k(V)$ satisfying $d(S, U) \le m$ one of the following possibilities is realized:

- if $d(S, U) = 1$ then $\{S, U\}^{mm}$ coincides with the line $[S \cap U, S + U]_k$,
- if $d(S, U) > 1$ then $\{S, U\}^{mm} = \{S, U\}$.

In other words, S and U are adjacent vertices of $\Gamma_k(V)$ if and only if the subset $\{S, U\}^{mm}$ contains more than 2 elements. Therefore, if f is a bijective transformation of $\mathcal{G}_k(V)$ such that

$$d(S, U) \le m \iff d(f(S), f(U)) \le m$$

for any $S, U \in \mathcal{G}_k(V)$ then f is an automorphism of $\Gamma_k(V)$.

Remark 3.5. In [Huang and Havlicek (2008); Huang (2010)] similar results were obtained for a class of graphs satisfying some technical conditions. These conditions hold for Grassmann graphs and we refer [Huang (2011)] for more examples of such graphs.

Consider the graphs $\Gamma_{i,j;m}(V)$ and $\Gamma_{i,j;\geq m}(V)$ whose vertex set is

$$\mathcal{G}_i(V) \cup \mathcal{G}_j(V)$$

and whose edges are pairs $S \in \mathcal{G}_i(V)$ and $U \in \mathcal{G}_j(V)$ satisfying

$$\dim(S \cap U) = m \quad \text{or} \quad \dim(S \cap U) \geq m,$$

respectively. Note that $\Gamma_{k,k;k-1}(V)$ is the Grassmann graph $\Gamma_k(V)$. If these graphs are non-trivial then their automorphisms are induced by automorphisms of V and, in some special cases, by isomorphisms of V to V^* [De Schepper and Van Maldeghem (2014)].

There is an analogue of Chow's theorem for twisted Grassmann graphs [Fujisaki, Koolen and Tagami (2006)]. As in Example 3.3, we suppose that $n = 2k+1$ and H is an $(n-1)$-dimensional subspace of V. Also, we assume that V is a vector space over a finite field. By Example 3.3, the associated twisted Grassmann graph $\tilde{\Gamma}_k(V, H)$ has precisely four distinct types of maximal cliques. An easy calculation shows that two maximal cliques are of the same type if they are of the same cardinality. The converse statement fails, since two maximal cliques of the first type $\mathcal{C}(S)$ and $\mathcal{C}(U)$ with $S, U \in \mathcal{G}_k(V)$ have distinct cardinalities if $S \subset H$ and $U \not\subset H$. Therefore, every automorphism f of $\tilde{\Gamma}_k(V, H)$ preserves the types of all maximal cliques. In particular, it sends every maximal clique $\mathcal{C}(S)$, $S \in \mathcal{G}_k(V)$ to a maximal clique of the same type. Two distinct maximal cliques $\mathcal{C}(S)$ and $\mathcal{C}(U)$ have a non-empty intersection if and only if S and U are adjacent vertices of $\Gamma_k(V)$. This implies that f induces an automorphism of $\Gamma_k(V)$. By Chow's theorem, this automorphism is induced by a semilinear automorphism l of V. The subspace H is invariant for l (recall that maximal cliques $\mathcal{C}(S)$ and $\mathcal{C}(U)$ have distinct cardinalities if $S \subset H$ and $U \not\subset H$). It is not difficult to prove that f is induced by l. So, every automorphism of $\tilde{\Gamma}_k(V, H)$ is induced by a semilinear automorphism of V preserving H.

Problem 3.3. Describe isometric embeddings of twisted Grassmann graphs.

3.10   Huang's theorem

As in the previous section, we suppose that V is an n-dimensional left vector space over a division ring and consider the Grassmann graph $\Gamma_k(V)$ such that $1 < k < n - 1$.

If V is a vector space over a finite field then $\mathcal{G}_k(V)$ is a finite set and it is clear that every bijective transformation of $\mathcal{G}_k(V)$ sending adjacent vertices of $\Gamma_k(V)$ to adjacent vertices of $\Gamma_k(V)$ is an automorphism of $\Gamma_k(V)$. The same holds for the general case.

Theorem 3.2 ([Huang (1998)]). *Every bijective transformation of the Grassmannian $\mathcal{G}_k(V)$ sending adjacent vertices of $\Gamma_k(V)$ to adjacent vertices of $\Gamma_k(V)$ is an automorphism of $\Gamma_k(V)$.*

Remark 3.6. In [Brauner (1988); Havlicek (1995)] this statement was proved for Grassmannians formed by 2-dimensional subspaces.

Theorem 3.2 is closely connected to Problem 1.1 and Kreuzer's example considered in Chapter 1.

Example 3.17. In Section 1.6 we constructed a semilinear 3-embedding of a vector space W whose dimension is greater than 3 (possibly infinite) in a 3-dimensional vector space W' which induces a bijection between the Grassmannians formed by 2-dimensional subspaces of W and W'. Consider the associated Grassmann graphs. In the second graph, any two distinct vertices are adjacent (the vector space W' is 3-dimensional). Thus our bijection sends adjacent vertices to adjacent vertices, but it is not an isomorphism between the Grassmann graphs.

Example 3.18. Let $l : V \to V'$ be the semilinear m-embedding described in Problem 1.1, i.e. V' is a vector space of dimension less than n and $(l)_{m-1}$ is a bijection of $\mathcal{G}_{m-1}(V)$ to $\mathcal{G}_{m-1}(V')$. Then $(l)_{m-1}$ transfers adjacent vertices of $\Gamma_{m-1}(V)$ to adjacent vertices of $\Gamma_{m-1}(V')$, but it is not an isomorphism between these graphs.

We start to prove Theorem 3.2. Let f be a bijective transformation of $\mathcal{G}_k(V)$ which sends adjacent vertices of $\Gamma_k(V)$ to adjacent vertices of $\Gamma_k(V)$. Then $\check{f}$ is a bijective transformation of $\mathcal{G}_{n-k}(V^*)$ and it maps adjacent vertices of $\Gamma_{n-k}(V^*)$ to adjacent vertices of $\Gamma_{n-k}(V^*)$. Also, $\check{f}$ is an automorphism of $\Gamma_{n-k}(V^*)$ if and only if f is an automorphism of $\Gamma_k(V)$. Therefore, we can restrict ourself to the case when $k \leq n - k$.

Note that the image of every maximal clique of $\Gamma_k(V)$ is contained in a maximal clique of $\Gamma_k(V)$. Also, we have

$$d(X, Y) \geq d(f(X), f(Y)) \quad \forall\, X, Y \in \mathcal{G}_k(V). \tag{3.18}$$

For every $X \in \mathcal{G}_k(V)$ and every subset $\mathcal{Y} \subset \mathcal{G}_k(V)$ we define the distance $d(X, \mathcal{Y})$ as the smallest distance $d(X, Y)$, where $Y \in \mathcal{Y}$.

3.10.1 *Proof of Theorem 3.2 for $n = 2k$*

Let $n = 2k$. Our first step is the following.

Lemma 3.13. *The mapping f transfers maximal cliques of $\Gamma_k(V)$ to maximal cliques of $\Gamma_k(V)$.*

Proof. Suppose that $\mathcal{M}$ and $\mathcal{M}'$ are maximal cliques of $\Gamma_k(V)$ such that $f(\mathcal{M})$ is contained in $\mathcal{M}'$. We need to show that $f(\mathcal{M}) = \mathcal{M}'$.

Let $X \in \mathcal{M}'$. First, we establish the existence of distinct

$$X_1, \ldots, X_k \in \mathcal{G}_k(V)$$

such that for every $i \in \{1, \ldots, k\}$ we have

$$d(X_i, \mathcal{M}') = d(X_i, X) = k - 1$$

and for every $Y \in \mathcal{M}' \setminus \{X\}$ there is at least one i such that $d(X_i, Y) = k$.

In the case when $\mathcal{M}'$ is a top $\langle U \rangle_k$, we take any $T \in \mathcal{G}_{k-1}(V)$ satisfying $U + T = V$ and 1-dimensional subspaces $P_1, \ldots, P_k$ spanning X. Then

$$X_i := T + P_i, \quad i \in \{1, \ldots, k\}$$

are as required. If $\mathcal{M}'$ is a star then we apply the above arguments to the top $\mathcal{M}'^0$ and use the fact that the annihilator mapping induces an isomorphism between $\Gamma_k(V)$ and $\Gamma_k(V^*)$.

Consider $Y_1, \ldots, Y_k \in \mathcal{G}_k(V)$ such that $f(Y_i) = X_i$ and define

$$\mathcal{H}_i := \{Y \in \mathcal{M} : d(Y_i, Y) = k - 1\},$$

$$\mathcal{H}'_i := \{Y \in \mathcal{M}' : d(X_i, Y) = k - 1\}$$

for every $i \in \{1, \ldots, k\}$. We have

$$d(Y_i, \mathcal{M}) \leq k - 1$$

(this is obvious if $\mathcal{M}$ is a star and this follows from the condition $n = 2k$ if $\mathcal{M}$ is a top). By (3.18),

$$k - 1 \geq d(Y_i, \mathcal{M}) \geq d(X_i, \mathcal{M}') = k - 1$$

and we get that

$$d(Y_i, \mathcal{M}) = k - 1.$$

Thus $\mathcal{H}_i$ is non-empty and the condition $d(X_i, \mathcal{M}') = k - 1$ together with
(3.18) imply that

$$f(\mathcal{H}_i) \subset \mathcal{H}'_i.$$

Then

$$f\left(\bigcap_{i=1}^{k} \mathcal{H}_i\right) \subset \bigcap_{i=1}^{k} f(\mathcal{H}_i) \subset \bigcap_{i=1}^{k} \mathcal{H}'_i = \{X\}.$$

The intersection of all $\mathcal{H}_i$ is non-empty. Indeed, if $\mathcal{M}$ is a top $\langle U \rangle_k$ then
every $Y_i \cap U$ is 1-dimensional (since $d(Y_i, \mathcal{M}) = k-1$) and there is $Y \in \langle U \rangle_k$
containing all $Y_i \cap U$, it belongs to every $\mathcal{H}_i$. In the case when $\mathcal{M}$ is a star, we
apply the above arguments to the top $\mathcal{M}^0$ and use the annihilator mapping.

So, for every $X \in \mathcal{M}'$ there exists $Y \in \mathcal{M}$ such that $f(Y) = X$ and we
get the claim.   □

Lemma 3.13 shows that f induces an injective transformation of the
vertex set of the graph $\mathrm{Cl}_k(V)$ (see Section 3.2). If two vertices of $\mathrm{Cl}_k(V)$ are
adjacent then the corresponding maximal cliques $\mathcal{M}$ and $\mathcal{M}'$ of $\Gamma_k(V)$ are
intersecting in a line. Then the intersection of the maximal cliques $f(\mathcal{M})$
and $f(\mathcal{M}')$ contains more than one element. The latter implies that this
intersection is a line and the corresponding vertices of $\mathrm{Cl}_k(V)$ are adjacent.
Hence our injection sends adjacent vertices of $\mathrm{Cl}_k(V)$ to adjacent vertices
of $\mathrm{Cl}_k(V)$. Then it transfers paths to paths and Lemma 3.2 guarantees that
two maximal cliques $\mathcal{M}$ and $\mathcal{M}'$ of $\Gamma_k(V)$ are of the same type if and only
if the maximal cliques $f(\mathcal{M})$ and $f(\mathcal{M}')$ are of the same type.

Since every line of $\mathcal{G}_k(V)$ is the intersection of two maximal cliques
of different types, f transfers lines to lines. If $\mathcal{M}$ is a maximal clique of
$\Gamma_k(V)$ then $\mathcal{M}$ and $f(\mathcal{M})$ (together with the lines contained in them) are
k-dimensional projective spaces. By Lemma 2.1, the restriction of f to $\mathcal{M}$
is a collineation between these projective spaces.

Now we suppose that X_1 and X_2 are vertices of $\Gamma_k(V)$ whose images
are adjacent vertices and show that X_1 and X_2 are adjacent.

We take two stars $\mathcal{S}_1$ and $\mathcal{S}_2$ containing X_1 and X_2, respectively. Then
$f(\mathcal{S}_1)$ and $f(\mathcal{S}_2)$ are maximal cliques of the same type. There is a unique
maximal clique $\mathcal{M}$ containing $f(X_1), f(X_2)$ and such that $\mathcal{M}$ and $f(\mathcal{S}_i)$
are of different types. This is the top

$$\langle f(X_1) + f(X_2) \rangle_k$$

if both $f(\mathcal{S}_i)$ are stars or the star

$$[f(X_1) \cap f(X_2)\rangle_k$$

if both $f(\mathcal{S}_i)$ are tops. Then $\mathcal{M} \cap f(\mathcal{S}_i)$ is a line and

$$f^{-1}(\mathcal{M} \cap f(\mathcal{S}_i))$$

is a line which contains X_i and is contained in $\mathcal{S}_i$. Denote by $\mathcal{T}_i$ the top containing this line. Then $f(\mathcal{T}_i)$ and $f(\mathcal{S}_i)$ are maximal cliques of different types. Thus $f(\mathcal{T}_i)$ and $\mathcal{M}$ are maximal cliques of the same type and their intersection contains a line. This is possible only in the case when $f(\mathcal{T}_i) = \mathcal{M}$. Therefore, $\mathcal{T}_1 = \mathcal{T}_2$. Since X_i belongs to $\mathcal{T}_i$, the vertices X_1 and X_2 are in the same maximal clique which implies that they are adjacent.

3.10.2   *Proof of Theorem 3.2 for $n \neq 2k$*

Now we suppose that $n \neq 2k$. It was noted above that we can restrict ourself to the case when $k < n - k$.

If $\mathcal{S}$ is a star and $\mathcal{T}$ is a top then

$$\max\{d(X, \mathcal{S}) : X \in \mathcal{G}_k(V)\} = k - 1, \quad \max\{d(X, \mathcal{T}) : X \in \mathcal{G}_k(V)\} = k$$

and (3.18) implies that $f(\mathcal{S})$ cannot be contained in a top. Therefore, f transfers stars to subsets of stars.

Lemma 3.14. *Let $\mathcal{S}_1$ and $\mathcal{S}_2$ be distinct stars with a non-empty intersection. Then $f(\mathcal{S}_1)$ and $f(\mathcal{S}_2)$ cannot be contained in the same star.*

Proof. Suppose that

$$\mathcal{S}_i = [S_i\rangle_k, \quad S_i \in \mathcal{G}_{k-1}(V), \quad i \in \{1, 2\}$$

and there is a star $\mathcal{S}$ containing both $f(\mathcal{S}_1)$ and $f(\mathcal{S}_2)$. Since the intersection of $\mathcal{S}_1$ and $\mathcal{S}_2$ is non-empty,

$$Q := S_1 \cap S_2 \in \mathcal{G}_{k-2}(V).$$

For every $X \in [Q\rangle_k$ there exist distinct $X_i \in \mathcal{S}_i$, $i \in \{1, 2\}$ such that

$$S_3 := X \cap X_1 \cap X_2 \in \mathcal{G}_{k-1}(V).$$

Then $f([S_3\rangle_k)$ is contained in a certain star $\mathcal{S}'$. The intersection of the stars $\mathcal{S}$ and $\mathcal{S}'$ contains $f(X_1)$ and $f(X_2)$ which implies that $\mathcal{S} = \mathcal{S}'$ and $f([S_3\rangle_k)$ is contained in $\mathcal{S}$.

Therefore, $f([Q\rangle_k)$ is a subset of $\mathcal{S}$. Since

$$\max\{d(X, [Q\rangle_k) : X \in \mathcal{G}_k(V)\} = k - 2$$

and

$$\max\{d(X, \mathcal{S}) : X \in \mathcal{G}_k(V)\} = k - 1,$$

the inclusion $f([Q\rangle_k) \subset \mathcal{S}$ contradicts (3.18). □

Let $\mathcal{T}$ be a top. We choose two distinct stars $\mathcal{S}_1$ and $\mathcal{S}_2$ such that $\mathcal{S}_1 \cap \mathcal{S}_2 \neq \emptyset$ and each $\mathcal{S}_i \cap \mathcal{T}$ is a line. Then $f(\mathcal{S}_1)$ and $f(\mathcal{S}_2)$ are subsets of some stars $\mathcal{S}'_1$ and $\mathcal{S}'_2$, respectively. If $f(\mathcal{T})$ is contained in a star $\mathcal{S}$ then every $\mathcal{S}'_i$ intersects $\mathcal{S}$ in a set containing at least two elements. Thus $\mathcal{S}'_1 = \mathcal{S} = \mathcal{S}'_2$ and every $f(\mathcal{S}_i)$ is contained in $\mathcal{S}$ which contradicts Lemma 3.14.

So, the images of tops cannot be contained in stars and f transfers tops to subsets of tops. It was established above that f sends stars to subsets of stars. Therefore, f maps lines to subsets of lines.

Lemma 3.15. *The mapping f transfers lines to lines.*

Proof. For every $S \in \mathcal{G}_{k-1}(V)$ there exists $S' \in \mathcal{G}_{k-1}(V)$ such that

$$f([S\rangle_k) \subset [S'\rangle_k.$$

The stars $[S\rangle_k$ and $[S'\rangle_k$ (together with the lines contained in them) are isomorphic to the projective spaces $\Pi_{V/S}$ and $\Pi_{V/S'}$, respectively. The restriction of f to $[S\rangle_k$ induces an injection of $\Pi_{V/S}$ to $\Pi_{V/S'}$ transferring lines to subsets of lines. The image of this mapping is not contained in a line of $\Pi_{V/S'}$, since $f([S\rangle_k)$ cannot be contained in a top. By Corollary 2.1, for every $U \in [S\rangle_{n-k+1}$ there exists $U' \in [S'\rangle_{n-k+1}$ such that

$$f([S, U]_k) \subset [S', U']_k.$$

We show that $f([S, U]_k)$ coincides with $[S', U']_k$.

Let $X' \in [S', U']_k$. We choose $Y' \in \mathcal{G}_k(V)$ such that

$$S' \cap Y' = 0 \quad \text{and} \quad X' \cap Y' = U' \cap Y'$$

is 1-dimensional (this is possible, since $\dim U' = n - k + 1$). Then

$$d(Y', X') = k - 1 \quad \text{and} \quad d(Y', Z') = k \quad \forall \, Z' \in [S', U']_k \setminus \{X'\}. \quad (3.19)$$

We take $Y \in \mathcal{G}_k(V)$ satisfying $f(Y) = Y'$. Since $\dim(U \cap Y) > 0$, there exists $X \in [S, U]_k$ such that $d(Y, X) \leq k - 1$. We have

$$k - 1 \geq d(Y, X) \geq d(Y', f(X)) \geq d(Y', [S', U']_k) = k - 1$$

which implies that

$$d(Y', f(X)) = k - 1.$$

By (3.19), this is possible only in the case when $f(X) = X'$. Therefore, for every $X' \in [S', U']_k$ there exists $X \in [S, U]_k$ such that $f(X) = X'$ and we get the claim.

Since

$$\dim U/S = \dim U'/S' = n - 2k + 2 \geq 3,$$

$[S, U]_k$ and $[S', U']_k$ (together with the lines contained in them) are projective spaces of the same finite dimension and the restriction of f to $[S, U]_k$ is a semicolliniation. By Corollary 2.3, this is a collineation.

The required statement follows from the fact that every line of $\mathcal{G}_k(V)$ is contained in a certain $[S, U]_k$, where $S \in \mathcal{G}_{k-1}(V)$ and $U \in [S\rangle_{n-k+1}$. □

Let $\mathcal{M}$ and $\mathcal{M}'$ be maximal cliques of $\Gamma_k(V)$ such that $f(\mathcal{M}) \subset \mathcal{M}'$. These maximal cliques are of the same type. Hence $\mathcal{M}$ and $\mathcal{M}'$ (together with the lines contained in them) are projective spaces of the same dimension. Lemmas 2.1 and 3.15 imply that $f(\mathcal{M})$ coincides with $\mathcal{M}'$.

So, f transfers maximal cliques of $\Gamma_k(V)$ to maximal cliques of $\Gamma_k(V)$. As in the case when $n = 2k$, we establish that f is an automorphism of $\Gamma_k(V)$.

Chapter 4

Johnson graph in Grassmann graph

The Johnson graph $J(n,k)$ consists of all k-element subsets of an n-element set. This is a thin version of the Grassmann graph $\Gamma_k(V)$, where V is an n-dimensional vector space. As in the Grassmann graph, any two distinct vertices of $J(n,k)$ are adjacent if $k = 1, n-1$ and we suppose that $1 < k < n-1$.

We investigate isometric embeddings of the Johnson graph $J(i,j)$ in the Grassmann graph $\Gamma_k(V)$ for all admissible pairs i, j. The images of such embeddings are said to be $J(i,j)$-subsets of the Grassmannian $\mathcal{G}_k(V)$. The description of these subsets is based on m-independent subsets of projective spaces introduced in Section 2.4 and we will use arguments similar to the arguments from the proof of Theorem 3.1.

If V is a vector space over a finite field then every $J(i,j)$-subset of $\mathcal{G}_k(V)$ corresponds to an equivalence class of linear codes. This equivalence class is defined by the associated m-independent subset, see Section 2.7. The classification of $J(i,j)$-subsets is equivalent to the classification of such linear codes.

Our description of $J(i,j)$-subsets is connected to the general problem of characterizing apartments in building Grassmannians (Section 4.5).

In Chapter 5 we give a characterization of isometric embeddings of Grassmann graphs in terms of $J(i,j)$-subsets.

4.1 Johnson graph

The *Johnson graph* $J(n,k)$ is formed by all k-element subsets of $\{1,\dots,n\}$. Two k-element subsets X and Y are adjacent vertices of this graph if

$$|X \cap Y| = k-1,$$

or equivalently,

$$|X \cup Y| = k + 1.$$

In the case when $k = 1, n - 1$, any two distinct vertices of $J(n, k)$ are adjacent. The graph $J(n, k)$ contains precisely

$$\binom{n}{k} = \frac{n!}{k!(n - k)!}$$

vertices.

For every subset $X \subset \{1, \dots, n\}$ we denote by X^c the complement of X in $\{1, \dots, n\}$. We need the following analogue of Proposition 3.1.

Proposition 4.1. *The complement mapping $X \to X^c$ induces an isomorphism between $J(n, k)$ and $J(n, n - k)$.*

Also, there is the direct analogue of Proposition 3.2.

Proposition 4.2. *The graph $J(n, k)$ is connected. If X and Y are vertices of $J(n, k)$ then the distance $d(X, Y)$ is equal to*

$$k \quad |X \cap Y| = |X \cup Y| - k.$$

The diameter of $J(n, k)$ is equal to $\min\{k, n - k\}$.

The proofs of these statements are similar to the proofs of Propositions 3.1 and 3.2 (respectively) and we leave them as exercises for the readers.

Remark 4.1. If X is a vertex of $J(2k, k)$ then X^c is the unique vertex of $J(2k, k)$ opposite to X, i.e. such that $d(X, X^c) = k$. Every vertex of $J(2k, k)$ belongs to a certain geodesic joining X and X^c. If $n \neq 2k$ then for every vertex of $J(n, k)$ there are a few opposite vertices.

It was noted above that any two distinct vertices of $J(n, k)$ are adjacent if $k = 1, n - 1$. In the case when $1 < k < n - 1$, maximal cliques of $J(n, k)$, as for Grassmann graphs, are stars and tops.

Example 4.1. Let X be a $(k - 1)$-element subset of $\{1, \dots, n\}$. The associated *star* is formed by all k-element subsets containing X. This is a clique of $J(n, k)$ consisting of $n - k + 1$ vertices.

Example 4.2. Let Y be a $(k + 1)$-element subset of $\{1, \dots, n\}$. The corresponding *top* is formed by all k-element subsets contained in Y. This is a clique of $J(n, k)$ consisting of $k + 1$ vertices.

The complement isomorphism of $J(n, k)$ to $J(n, n - k)$ transfers stars to tops and tops to stars. The proof of the following statement is similar to the proof of Proposition 3.3.

Proposition 4.3. *If $1 < k < n - 1$ then every maximal clique of $J(n, k)$ is a star or a top.*

As above, we suppose that $1 < k < n - 1$. The intersection of two distinct maximal cliques of $J(n, k)$ contains at most two vertices. This intersection is maximal, i.e. it contains precisely two vertices, if and only if the cliques are of different types and the corresponding $(k - 1)$-element and $(k + 1)$-element subsets are incident (one of them is contained in the other). The intersection of two distinct stars is non-empty if and only if the associated $(k - 1)$-element subsets are adjacent vertices of $J(n, k - 1)$. Similarly, the intersection of two distinct tops is non-empty if and only if the associated $(k + 1)$-element subsets are adjacent vertices of $J(n, k + 1)$.

Consider the graph $\mathrm{Cl}(n, k)$ whose vertex set consists of all subsets of $\{1, \ldots, n\}$ containing $k - 1$ or $k + 1$ elements and whose edges are pairs X, Y such that X is a $(k - 1)$-element subset contained in a $(k + 1)$-element subset Y. In other words, the vertex set of $\mathrm{Cl}(n, k)$ is formed by all maximal cliques of $J(n, k)$ and two maximal cliques are adjacent vertices if their intersection is maximal. As in Section 3.2, we prove the following.

Lemma 4.1. *The graph $\mathrm{Cl}(n, k)$ is connected. If $X_0, \ldots, X_i$ is a path in $\mathrm{Cl}(n, k)$ then $|X_0| = |X_i|$ only in the case when i is even.*

Every permutation on the set $\{1, \ldots, n\}$ induces an automorphism of $J(n, k)$. The complement mapping is an automorphism of $J(n, k)$ if $n = 2k$. If s is a permutation on $\{1, \ldots, n\}$ then $s(X^c) = s(X)^c$ for every subset $X \subset \{1, \ldots, n\}$. Therefore, for $n = 2k$ the automorphisms of $J(n, k)$ induced by permutations are commuting with the complement mapping.

There is the following analogue of Chow's theorem.

Theorem 4.1. *If $n \neq 2k$ then every automorphism of $J(n, k)$ is induced by a permutation on $\{1, \ldots, n\}$. In the case when $n = 2k$, every automorphism of $J(n, k)$ is induced by a permutation on $\{1, \ldots, n\}$ or it is the composition of the complement mapping and the automorphism induced by a permutation.*

Proof. Let f be an automorphism of $J(n, k)$. The mapping

$$X \to f(X^c)^c$$

is an automorphism of $J(n, n - k)$. If it is induced by a permutation s on $\{1, \dots, n\}$ then for every vertex X of $J(n, k)$ we have

$$f(X) = s(X^c)^c = s(X)$$

which means that f is induced by s. Thus we can restrict ourself to the case when $1 \leq k \leq n - k$.

For $k = 1$ the statement is trivial and we suppose that $1 < k \leq n - k$. It is clear that f transfers maximal cliques to maximal cliques. If $k < n - k$ then stars and tops have distinct cardinalities which guarantees that stars go to stars and tops go to tops. Hence f induces a bijective transformation of the vertex set of $J(n, k - 1)$. This is an automorphism of $J(n, k - 1)$, since two distinct stars of $J(n, k)$ have a non-empty intersection if and only if the associated $(k-1)$-element subsets are adjacent vertices of $J(n, k-1)$. Step by step, we come to an automorphism of $J(n, 1)$, i.e. a permutation on $\{1, \dots, n\}$. This permutation induces f.

Suppose that $n = 2k$. In this case, stars and tops are of the same cardinality. It is easy to see that f induces an automorphism of $\mathrm{Cl}(n, k)$. Using Lemma 4.1, we show that one of the following possibilities is realized:

- stars go to stars and tops go to tops,
- stars go to tops and tops go to stars.

It follows from the above arguments that f is induced by a permutation on $\{1, \dots, n\}$ in the first case. In the second case, we consider the composition of f and the complement mapping. This automorphism sends stars to stars and tops to tops. Hence it is induced by a permutation on $\{1, \dots, n\}$. Then f is the composition of the complement mapping and the automorphism induced by a permutation. $\qquad\square$

Remark 4.2. By Theorem 4.1, the automorphism group of $J(n, k)$ is isomorphic to the permutation group S_n if $n \neq 2k$. In the case when $n = 2k$, this group is the direct product of S_n and $\mathbb{Z}_2$.

Remark 4.3. The classical proof of Chow's theorem is similar to the proof given above and based on the Fundamental Theorem of Projective Geometry.

Remark 4.4. We refer [Brouwer, Cohen and Neumaier (1989), Section 9.1] for more information concerning Johnson graphs.

4.2 Isometric embeddings of Johnson graphs in Grassmann graphs

Let V be a left vector space over a division ring. Suppose that $\dim V = n$ is finite. We investigate isometric embeddings of the Johnson graph $J(i, j)$ in the Grassmann graph $\Gamma_k(V)$. The following facts concerning embeddings of $J(i, j)$ in $\Gamma_k(V)$ (not necessarily isometric) are obvious:

- if $j = 1, i - 1$ then every embedding of $J(i, j)$ in $\Gamma_k(V)$ is an injection to a clique of $\Gamma_k(V)$,
- there are no embeddings of $J(i, j)$ in $\Gamma_k(V)$ if $1 < j < i - 1$ and $k = 1, n - 1$.

For this reason, we will suppose that

$$1 < j < i - 1 \text{ and } 1 < k < n - 1.$$

These conditions shows that n and i both are not less than 4. The existence of isometric embeddings of $J(i, j)$ in $\Gamma_k(V)$ implies that

$$\min\{j, i - j\} \leq \min\{k, n - k\}, \qquad (4.1)$$

i.e. the diameter of $J(i, j)$ is not greater than the diameter of $\Gamma_k(V)$. In the case when $j = 2, i - 2$, the diameter of $J(i, j)$ is equal to 2 and every embedding of $J(i, j)$ is isometric.

The images of isometric embeddings of $J(i, j)$ in $\Gamma_k(V)$ will be called $J(i, j)$-*subsets* of $\mathcal{G}_k(V)$. Since the Johnson graphs $J(i, j)$ and $J(i, i - j)$ are isomorphic, $J(i, j)$-subsets and $J(i, i - j)$-subsets are coincident and we can restrict ourself to the case when $1 < j \leq i - j$.

Example 4.3. Let $B = \{x_1, \ldots, x_n\}$ be a base of V. Consider the subset of $\mathcal{G}_k(V)$ consisting of all k-dimensional subspaces spanned by subsets of B, i.e. all subspaces of type

$$\langle x_{i_1}, \ldots, x_{i_k} \rangle,$$

where $i_1, \ldots, i_k$ are mutually distinct. This subset is known as the *apartment* of $\mathcal{G}_k(V)$ associated to the base B (the general concept of apartments in building Grassmannians will be discussed in Section 4.5). This is a $J(n, k)$-subset. It follows from Lemma 1.7 that the annihilator mapping transfers our apartment to the apartment of $\mathcal{G}_{n-k}(V^*)$ defined by the base of V^* dual to B. Note that two bases of V define the same apartment if and only if the vectors from one of the bases are scalar multiples of the vectors from the other, in other words, these bases define the same base of the projective space Π_V.

Remark 4.5. It is trivial that a subset of $\mathcal{G}_1(V)$ is an apartment if and only if this subset is a base of Π_V. Also, every apartment of $\mathcal{G}_{n-1}(V)$ is a base of Π_V^*. Conversely, if $U_1, \ldots, U_n \in \mathcal{G}_{n-1}(V)$ form a base of Π_V^* then their annihilators form a base of Π_{V^*} and we take any of the associated bases of V^*. Lemma 1.7 shows that $U_1, \ldots, U_n$ form the apartment of $\mathcal{G}_{n-1}(V)$ defined by the dual base of V. So, a subset of $\mathcal{G}_{n-1}(V)$ is an apartment if and only if this subset is a base of Π_V^*.

Let $\mathcal{X} \subset \mathcal{G}_1(V)$ be an m-independent subset of Π_V. This means that $\mathcal{X}$ contains at least m elements and every m-element subset of $\mathcal{X}$ is independent. If $k \in \{2, \ldots, m\}$ and $P_1, \ldots, P_k$ are distinct elements of $\mathcal{X}$ then $P_1 + \cdots + P_k$ is a k-dimensional subspace of V. The set formed by all such k-dimensional subspaces is denoted by $J_k(\mathcal{X})$. If $\mathcal{X}$ is independent then $J_k(\mathcal{X})$ is an apartment of $\mathcal{G}_k(U)$, where U is the sum of all elements from $\mathcal{X}$. In particular, $J_k(\mathcal{X})$ is an apartment of $\mathcal{G}_k(V)$ if $\mathcal{X}$ is a base of Π_V.

Proposition 4.4. *Suppose that $\mathcal{X} \subset \mathcal{G}_1(V)$ is an m-independent subset of Π_V consisting of l elements and $k \in \{2, \ldots, m\}$. The following assertions are fulfilled:*

(1) *if $k \leq m - 2$ then $J_k(\mathcal{X})$ is the image of an embedding of $J(l, k)$ in $\Gamma_k(V)$,*

(2) *if $2k \leq m$ then $J_k(\mathcal{X})$ is a $J(l, k)$-subset.*

Proof. Let $P_1, \ldots, P_l$ be the elements of $\mathcal{X}$. For every subset $I \subset \{1, \ldots, l\}$ we define

$$S_I := \sum_{i \in I} P_i.$$

Since $\mathcal{X}$ is m-independent, we have

$$\dim S_I = |I| \quad \text{if} \quad |I| \leq m.$$

Consider the mapping which transfers every k-element subset $I \subset \{1, \ldots, l\}$ to the k-dimensional subspace S_I.

(1). Let $k \leq m - 2$. We show that S_I and S_J are adjacent vertices of $\Gamma_k(V)$ if and only if I and J are adjacent vertices of $J(l, k)$. If I and J are adjacent then

$$|I \cup J| = k + 1 < m$$

which guarantees that the subspace $S_{I \cup J} = S_I + S_J$ is $(k+1)$-dimensional. If I and J are not adjacent in $J(n, k)$ then

$$|I \cup J| \geq k + 2.$$

If M is a $(k+2)$-element subset of $I \cup J$ then

$$\dim(S_I + S_J) \geq \dim S_M = k + 2,$$

which means that S_I and S_J are not adjacent in $\Gamma_k(V)$.

(2). Suppose that $2k \leq m$. If I and J are k-element subsets of $\{1, \ldots, l\}$ then $I \cup J$ contains at most $2k$ elements and we have

$$\dim(S_I + S_J) = |I \cup J|$$

which implies that $d(I, J) = d(S_I, S_J)$. $\qquad\square$

Remark 4.6. The first example of a $J(n,k)$-subset which is not an apartment of $\mathcal{G}_k(V)$ was constructed in [Cooperstein and Shult (1997)]. This is $J_2(\mathcal{X})$, where $n = 5$ and $\mathcal{X}$ is a 4-simplex in Π_V.

The following example shows that there exist non-isometric embeddings of Johnson graphs in Grassmann graphs.

Example 4.4. Suppose that $\mathcal{X} \subset \mathcal{G}_1(V)$ is an $(n-1)$-simplex and $n = 2k$ is not less than 6. Then $k \geq 3$. Since $\mathcal{X}$ is $(n-1)$-independent and

$$k = n - k \leq n - 3 = (n-1) - 2,$$

the first part of Proposition 4.4 shows that $J_k(\mathcal{X})$ is the image of an embedding of $J(n,k)$ in $\Gamma_k(V)$. This embedding is not isometric. Indeed, all elements of $J_k(\mathcal{X})$ are contained in a certain $(n-1)$-dimensional subspace of V which means that the distance between any two elements of $J_k(\mathcal{X})$ in $\Gamma_k(V)$ is less than k. On the other hand, the diameter of $J(n,k)$ is equal to k.

As in Proposition 4.4, we suppose that $\mathcal{X} \subset \mathcal{G}_1(V)$ is an m-independent subset of Π_V consisting of l elements and $k \in \{2, \ldots, m\}$. Denote by $J_k^*(\mathcal{X})$ the set formed by the annihilators of all elements from $J_k(\mathcal{X})$, i.e. we write $J_k^*(\mathcal{X})$ instead of $(J_k(\mathcal{X}))^0$. It is clear that $\mathcal{X}^0 \subset \mathcal{G}_{n-1}(V^*)$ is an m-independent subset of Π_{V^*}. The intersection of any k distinct elements of $\mathcal{X}^0$ is an $(n-k)$-dimensional subspace of V^* and the set of all such subspaces coincides with $J_k^*(\mathcal{X})$. It follows from Proposition 4.4 that $J_k^*(\mathcal{X})$ is a $J(l,k)$-subset of $\mathcal{G}_{n-k}(V^*)$ if $2k \leq m$.

By Example 4.3, $J_k^*(\mathcal{X})$ is an apartment of $\mathcal{G}_{n-k}(V^*)$ if $\mathcal{X}$ is a base of Π_V. Suppose that $\mathcal{X}$ is an independent subset of Π_V. Then $\mathcal{X}^0$ is an independent subset of Π_{V^*} and the intersection of all elements from $\mathcal{X}^0$ is a certain $(n-l)$-dimensional subspace $S \subset V^*$. If $U_1, \ldots, U_l$ are the elements of $\mathcal{X}^0$ then

$$U_1/S, \ldots, U_l/S$$

form a base of $\Pi^*_{V^*/S}$. By Remark 4.5, this is an apartment of $\mathcal{G}_{l-1}(V^*/S)$. If B is one of the associated bases of V^*/S then

$$J^*_k(\mathcal{X}) = \Phi_S(\mathcal{A}),$$

where $\mathcal{A}$ is the apartment of $\mathcal{G}_{l-k}(V^*/S)$ defined by B. Recall that for any subspaces $S, U \subset V$ such that $S \subset U$ the mapping Φ^U_S transfers every subspace of U/S to the corresponding subspace of V; if $U = V$ or $S = 0$ then this mapping is denoted by Φ_S or Φ^U, respectively.

Now we present two examples of $J(i,j)$-subsets of $\mathcal{G}_k(V)$. Since we assume that $1 < j \le i - j$, the condition (4.1) implies that

$$j \le \min\{i - j, k, n - k\}. \tag{4.2}$$

Example 4.5. Let $S \in \mathcal{G}_{k-j}(V)$. By (4.2), we have

$$\dim(V/S) = n - k + j \ge 2j.$$

Suppose that $\mathcal{X} \subset \mathcal{G}_1(V/S)$ is an m-independent subset of $\Pi_{V/S}$ consisting of i elements and $m \ge 2j$. Then $J_j(\mathcal{X})$ is a $J(i,j)$-subset of $\mathcal{G}_j(V/S)$ and

$$\Phi_S(J_j(\mathcal{X}))$$

is a $J(i,j)$-subset of $\mathcal{G}_k(V)$. If $i = 2j$ then $\mathcal{X}$ is an independent subset of $\Pi_{V/S}$ and $J_j(\mathcal{X})$ is an apartment of $\mathcal{G}_j(U/S)$, where $U \in \mathcal{G}_{k+j}(V)$.

Example 4.6. The condition (4.2) shows that $k+j \le n$. Let $U \in \mathcal{G}_{k+j}(V)$. Suppose that $\mathcal{Y} \subset \mathcal{G}_1(U^*)$ is an m-independent subset of Π_{U^*} consisting of i elements and $m \ge 2j$. Then $J_j(\mathcal{Y})$ is a $J(i,j)$-subset of $\mathcal{G}_j(U^*)$, $J^*_j(\mathcal{Y})$ is a $J(i,j)$-subset of $\mathcal{G}_k(U)$ and

$$\Phi^U(J^*_j(\mathcal{Y}))$$

is a $J(i,j)$-subset of $\mathcal{G}_k(V)$. If $i = 2j$ then $\mathcal{Y}$ is an independent subset and we have

$$J^*_j(\mathcal{Y}) = \Phi_S(\mathcal{A}),$$

where $S \in \mathcal{G}_{k-j}(U)$ and $\mathcal{A}$ is an apartment of $\mathcal{G}_j(U/S)$.

Theorem 4.2 ([Pankov 1 (2011)]). *Suppose that $\mathcal{J}$ is a $J(i,j)$-subset of $\mathcal{G}_k(V)$ and $1 < j \le i - j$. In the case when $i = 2j$, there exist*

$$S \in \mathcal{G}_{k-j}(V) \quad and \quad U \in \mathcal{G}_{k+j}(V)$$

such that

$$\mathcal{J} = \Phi^U_S(\mathcal{A}),$$

where $\mathcal{A}$ is an apartment of $\mathcal{G}_j(U/S)$. If $i \ne 2j$ then one of the following possibilities is realized:

(A) *there is $S \in \mathcal{G}_{k-j}(V)$ such that*
$$\mathcal{J} = \Phi_S(J_j(\mathcal{X})),$$
where $\mathcal{X} \subset \mathcal{G}_1(V/S)$ is an m-independent subset of $\Pi_{V/S}$ consisting of i elements and $m \geq 2j$;

(B) *there is $U \in \mathcal{G}_{k+j}(V)$ such that*
$$\mathcal{J} = \Phi^U(J_j^*(\mathcal{Y})),$$
where $\mathcal{Y} \subset \mathcal{G}_1(U^*)$ is an m-independent subset of Π_{U^*} consisting of i elements and $m \geq 2j$.

In the case when $i \neq 2j$, we say that $\mathcal{J}$ is a $J(i,j)$-*subset of type* (A) or a $J(i,j)$-*subset of type* (B) if the corresponding possibility is realized. If $i = 2j$ then all $J(i,j)$-subsets are of the same type.

Let $\mathcal{C}$ be a maximal clique of $\Gamma_k(V)$ (a star or a top). Let also $\mathcal{J}$ be a $J(i,j)$-subset of $\mathcal{G}_k(V)$. Suppose that the intersection of $\mathcal{C}$ and $\mathcal{J}$ contains more than one vertex. We say that this intersection is a *star* or a *top* of $\mathcal{J}$ if $\mathcal{C}$ is a star or a top, respectively.

Lemma 4.2. *Suppose that $i \neq 2j$. If $\mathcal{J}$ is a $J(i,j)$-subset of type* (A) *then the stars and the tops of $\mathcal{J}$ consist of $i-j+1$ and $j+1$ vertices, respectively. In the case when $\mathcal{J}$ is a $J(i,j)$-subset of type* (B), *the stars and the tops of $\mathcal{J}$ consist of $j+1$ and $i-j+1$ vertices, respectively.*

Proof. The first statement is obvious. The annihilator mapping sends $J(i,j)$-subsets of type (A) to $J(i,j)$-subsets of type (B), stars go to tops and tops go to stars. This implies the second statement. $\square$

Lemma 4.2 shows that the class of $J(i,j)$-subsets of type (A) and the class of $J(i,j)$-subsets of type (B) are disjoint, i.e. there is no $J(i,j)$-subset belonging to both these classes.

It follows from Theorem 4.2 that every $J(n,k)$-subset of $\mathcal{G}_k(V)$ is an apartment only in the case when $n = 2k$ (this fact is related to the property of $J(2k,k)$ described in Remark 4.1). If $k < n-k$ then apartments of $\mathcal{G}_k(V)$ are $J(n,k)$-subsets of type (A) defined by bases of the projective space Π_V. In the case when $k > n-k$, apartments of $\mathcal{G}_k(V)$ can be considered as $J(n,n-k)$-subsets of type (B) defined by bases of Π_{V^*}.

Remark 4.7. By [Cooperstein, Kasikova and Shult (2005)], every apartment can be characterized as a subset $\mathcal{A} \subset \mathcal{G}_k(V)$ such that the restriction of the Grassmann graph $\Gamma_k(V)$ to $\mathcal{A}$ is isomorphic to $J(n,k)$ and every maximal clique of this restriction is an independent subset of the corresponding projective space.

4.3   Proof of Theorem 4.2

The proof is similar to the proof of Theorem 3.1.

Lemma 4.3. *For every embedding of $J(i,j)$ in $\Gamma_k(V)$ the following assertions are fulfilled:*

(1) *Every maximal clique of $\Gamma_k(V)$ contains at most one image of a maximal clique of $J(i,j)$.*

(2) *The image of every maximal clique of $J(i,j)$ is contained in precisely one maximal clique of $\Gamma_k(V)$.*

Proof. Similar to the proof of Lemma 3.3.   □

Using Lemmas 4.1 and 4.3, we establish the direct analogue of Proposition 3.5.

Proposition 4.5. *For every embedding of $J(i,j)$ in $\Gamma_k(V)$ one of the following possibilities is realized:*

(A) *stars go to subsets of stars and tops go to subsets of tops,*

(B) *stars go to subsets of tops and tops go to subsets of stars.*

Let f be an isometric embedding of $J(i,j)$ in $\Gamma_k(V)$.

First we consider the case when f is an embedding of type (A). Since stars of $J(i,j)$ and $\Gamma_k(V)$ can be identified with vertices of the graphs $J(i,j-1)$ and $\Gamma_{k-1}(V)$ (respectively), f induces a mapping f_{j-1} of the vertex set of $J(i,j-1)$ to $\mathcal{G}_{k-1}(V)$. The following property is obvious: if X is a vertex of $J(i,j)$ and Y is a vertex of $J(i,j-1)$ then

$$Y \subset X \implies f_{j-1}(Y) \subset f(X).$$

Lemma 4.4. *The mapping f_{j-1} is an isometric embedding of $J(i,j-1)$ in $\Gamma_{k-1}(V)$. This embedding is of type (A).*

Proof. Similar to the proof of Lemma 3.6.   □

As in Section 3.5, we construct a sequence of isometric embeddings f_p of $J(i,p)$ in $\Gamma_{k-j+p}(V)$ with $p = j, \dots, 1$ and such that $f_j = f$. If X is a vertex of $J(i,p)$ and Y is a vertex of $J(i,p-1)$ then

$$Y \subset X \implies f_{p-1}(Y) \subset f_p(X)$$

for every $p \geq 2$. The vertices of the graph $J(i,1)$ are the one-element subsets $\{1\}, \dots, \{i\}$ and

$$Q_1 := f_1(\{1\}), \dots, Q_i := f_1(\{i\})$$

are distinct elements of $\mathcal{G}_{k-j+1}(V)$.

Lemma 4.5. *If $p \in \{2, \ldots, j\}$ then for any distinct $i_1, \ldots, i_p \in \{1, \ldots, n\}$ we have*

$$f_p(\{i_1, \ldots, i_p\}) = Q_{i_1} + \cdots + Q_{i_p}.$$

Proof. It is easy to see that $Q_{i_1}, \ldots, Q_{i_p}$ are contained in $f_p(\{i_1, \ldots, i_p\})$. As in the proof of Lemma 3.7, we use induction to show that there is no proper subspace of $f_p(\{i_1, \ldots, i_p\})$ containing $Q_{i_1}, \ldots, Q_{i_p}$. $\qquad\square$

Since f_1 is an isometric embedding of $J(i, 1)$ in $\Gamma_{k-j+1}(V)$, the vertices $Q_1, \ldots, Q_i$ form a clique in $\Gamma_{k-j+1}(V)$. As in the proof of Theorem 3.1, we establish that this clique is contained in a certain star

$$[S\rangle_{k-j+1}, \quad S \in \mathcal{G}_{k-j}(V).$$

Then

$$f = \Phi_S f',$$

where f' is an isometric embedding of $J(i, j)$ in $\Gamma_j(V/S)$. Denote by $\mathcal{X}$ the subset of $\mathcal{G}_1(V/S)$ formed by

$$P_1 := Q_1/S, \ldots, P_i := Q_i/S.$$

It follows from Lemma 4.5 that $\mathcal{X}$ is a j-independent subset of $\Pi_{V/S}$ and the image of f' coincides with $J_j(\mathcal{X})$.

Let $\mathcal{Y}$ be a $(2j)$-element subset of $\mathcal{X}$. Suppose that its elements are

$$P_{i_1}, \ldots, P_{i_j} \quad \text{and} \quad P_{k_1}, \ldots, P_{k_j}.$$

By (4.2), we have

$$\dim(V/S) = n - k + j \geq 2j.$$

Since the distance between $\{i_1, \ldots, i_j\}$ and $\{k_1, \ldots, k_j\}$ in $J(i, j)$ is equal to j, the distance between

$$f'(\{i_1, \ldots, i_j\}) = P_{i_1} + \cdots + P_{i_j} \quad \text{and} \quad f'(\{k_1, \ldots, k_j\}) = P_{k_1} + \cdots + P_{k_j}$$

in $\Gamma_j(V/S)$ is equal to j. Then the sum of these subspaces is $(2j)$-dimensional which implies that $\mathcal{Y}$ is an independent subset of $\Pi_{V/S}$.

So, the image of f coincides with

$$\Phi_S(J_j(\mathcal{X}))$$

and $\mathcal{X}$ is an m-independent subset of $\Pi_{V/S}$ such that $m \geq 2j$. If $i = 2j$ then $J_j(\mathcal{X})$ is an apartment of $\mathcal{G}_j(U/S)$ for a certain $U \in \mathcal{G}_{k+j}(V)$, see Example 4.5.

Consider the case when f is an embedding of type (B). Denote by f^* the composition of f and the annihilator mapping of $\mathcal{G}_k(V)$ to $\mathcal{G}_{n-k}(V^*)$. This is an isometric embedding of $J(i,j)$ in $\Gamma_{n-k}(V^*)$ satisfying (A). Hence its image is

$$\Phi_{S'}(J_j(\mathcal{X}')),$$

where $S' \in \mathcal{G}_{n-k-j}(V^*)$ and $\mathcal{X}'$ is an m-independent subset of $\Pi_{V^*/S'}$ consisting of i elements and such that $m \geq 2j$. We define

$$U := (S')^0 \in \mathcal{G}_{k+j}(V)$$

and observe that every element of $(\Phi_{S'}(\mathcal{X}'))^0$ is a $(k+j-1)$-dimensional subspace of U. It is clear that $(\Phi_{S'}(\mathcal{X}'))^0$ is an m-independent subset of Π_U^*. The intersection of any j distinct elements of $(\Phi_{S'}(\mathcal{X}'))^0$ is a k-dimensional subspace of U. The set of all such subspaces coincides with the image of f. Therefore, the image of f is

$$\Phi^U(J_j^*(\mathcal{Y})),$$

where $\mathcal{Y} \subset \mathcal{G}_1(U^*)$ is formed by the annihilators of the elements from $(\Phi_{S'}(\mathcal{X}'))^0$ in U^*. The latter is an m independent subset of Π_{U^*}. If $i = 2j$ then, by Example 4.6, $J_j^*(\mathcal{Y})$ coincides with $\Phi_S(\mathcal{A})$, where $S \in \mathcal{G}_{k-j}(U)$ and $\mathcal{A}$ is an apartment of $\mathcal{G}_j(U/S)$.

4.4   Classification problem and relations to linear codes

Recall that two subsets of $\mathcal{G}_k(V)$ are equivalent if there is an element of $\mathrm{P\Gamma L}(V)$ (the transformation of $\mathcal{G}_k(V)$ induced by a semilinear automorphism of V) which sends one of these subsets to the other.

It is clear that all $J(i,j)$-subsets of $\mathcal{G}_k(V)$ are equivalent if $i = 2j$. In the case when $i \neq 2j$, the classification of $J(i,j)$-subsets can be reduced to the classification of the associated m-independent subsets. It is easy to see that two $J(i,j)$-subsets

$$\Phi_S(J_j(\mathcal{X})) \quad \text{and} \quad \Phi_{S'}(J_j(\mathcal{X}'))$$

are equivalent if and only if there is a semilinear isomorphism of V/S to V/S' sending $\mathcal{X}$ to $\mathcal{X}'$. The same holds for $J(i,j)$-subsets of type (B).

Using Proposition 2.4 we describe one interesting class of $J(i,j)$-subsets.

Corollary 4.1. *Let $\mathcal{J}$ be a $J(i,j)$-subset of $\mathcal{G}_k(V)$ and $i \neq 2j$. The restriction of the Grassmann graph $\Gamma_k(V)$ to $\mathcal{J}$ is isomorphic to the Johnson* *graph $J(i,j)$. Suppose that every automorphism of this restriction can be extended to the automorphism of $\Gamma_k(V)$ induced by a linear automorphism of V, i.e. to an element of $\mathrm{PGL}(V)$. In the case when R is a field, the following assertions are fulfilled:*

(1) *If $\mathcal{J}$ is of type (A) then there is $S \in \mathcal{G}_{k-j}(V)$ such that*
$$\mathcal{J} = \Phi_S(J_j(\mathcal{X})),$$
where $\mathcal{X} \subset \mathcal{G}_1(V/S)$ is an independent subset of $\Pi_{V/S}$ consisting of i elements or an $(i-1)$-simplex.

(2) *If $\mathcal{J}$ is of type (B) then there is $U \in \mathcal{G}_{k+j}(V)$ such that*
$$\mathcal{J} = \Phi^U(J_j^*(\mathcal{Y})),$$
where $\mathcal{Y} \subset \mathcal{G}_1(U^*)$ is an independent subset of Π_{U^*} consisting of i elements or an $(i-1)$-simplex.

Proof. Suppose that $\mathcal{J} = \Phi_S(J_j(\mathcal{X}))$ is a $J(i,j)$-subset of type (A). The restriction of $\Gamma_j(V/S)$ to $J_j(\mathcal{X})$ is isomorphic to $J(i,j)$. By Theorem 4.1, every automorphism of this subgraph is induced by a permutation on $\mathcal{X}$. On the other hand, every automorphism of this subgraph can be extended to an element of $\mathrm{PGL}(V/S)$. The latter follows from our assumption and the fact that every linear automorphism of V preserving S induces a linear automorphism of V/S. Therefore, $\mathcal{X}$ is a PGL-subset and Proposition 2.4 gives the claim. In the case when $\mathcal{J}$ is a $J(i,j)$-subset of type (B), the proof is similar. $\square$

Suppose that V is a vector space over the finite field consisting of q elements. Consider any $J(i,j)$-subset $\Phi_S(J_j(\mathcal{X}))$ of type (A). By Section 2.7, the subset $\mathcal{X}$ defines an equivalence class of linear $[i,t]_q$ codes, where t is the dimension of the subspace of V/S spanned by $\mathcal{X}$ and $2j \le t \le i$. Note that if $\mathcal{X}$ is t-independent then the minimal weight of the corresponding codes is equal to $n - t + 1$ (Section 2.7). Similarly, every $J(i,j)$-subset of type (B) is related to a certain equivalence class of linear codes. Two $J(i,j)$-subsets of the same type are equivalent if and only if they correspond to the same equivalence class of linear codes.

4.5 Characterizations of apartments in building Grassmannians

We start from a very brief and informal description of buildings [Tits (1974)] and associated building Grassmannians (called also shadow spaces). We refer [Buekenhout and Cohen (2013); Pankov (2010); Pasini (1994); Shult (2011)] for more information.

Let X be a set (possibly infinite). A *simplicial complex* over X is a set formed by finite subsets of X called *simplices* and such that the following two conditions hold:

- every one-element subset is a simplex,
- every subset of a simplex is a simplex.

Elements of X are said to be *vertices* of this simplicial complex. For example, for every finite-dimensional vector space V the *flag complex* $\Delta(V)$ is the simplicial complex whose vertex set is formed by all proper subspaces of V and whose simplices are flags, i.e. chains of mutually distinct incident subspaces.

By [Tits (1974)], a *building* is a simplicial complex Δ together with a distinguished family of subcomplexes, so-called *apartments*, which satisfy some axioms. One of the axioms says that all apartments are isomorphic to a certain Coxeter complex, i.e. the simplicial complex obtained from a Coxeter system. This Coxeter system defines the type of the building. Maximal simplices of Δ are said to be *chambers*. They have the same cardinality n called the *rank* of the building. We say that two chambers C and C' are *adjacent* if

$$|C \cap C'| = n - 1.$$

The building Δ admits a *labeling*, i.e. a mapping of the vertex set to the set $\{1, \ldots, n\}$ whose restriction to every chamber is bijective. This labeling is unique up to a permutation on $\{1, \ldots, n\}$. Every subset in the vertex set consisting of all vertices labeled by the same integer is called a *Grassmannian* of Δ. So, the vertex set of Δ can be presented as the disjoint union of the building Grassmannians and this decomposition does not depend on a labeling. The intersections of apartments with a Grassmannian are said to be *apartments* in this Grassmannian. The *Grassmann graph* associated to a Grassmannian $\mathcal{G}$ is the graph whose vertex set is $\mathcal{G}$ and two vertices $x, y \in \mathcal{G}$ are adjacent if there is a simplex $S \in \Delta$ such that $S \cup \{x\}$ and $S \cup \{y\}$ are chambers (these chambers are adjacent).

The Coxeter system of type A_{n-1} is the pair (W, S), where $W = S_n$ is the permutation group on the set $\{1, \ldots, n\}$ and S is the set of all transpositions $(i, i + 1)$. Every building of type A_{n-1}, $n \geq 4$ is the flag complex $\Delta(V)$ of a certain n-dimensional vector space V over a division ring. The associated Grassmannians are $\mathcal{G}_k(V)$, $k \in \{1, \ldots, n-1\}$ whose Grassmann graphs are $\Gamma_k(V)$. Every apartment of $\Delta(V)$ is defined by a certain base B of V, it consists of all flags formed by the subspaces spanned by subsets of B. The associated apartment of $\mathcal{G}_k(V)$ is the set of all k-dimensional subspaces spanned by subsets of B, i.e. we come to the notion of an apartment considered in Section 4.2. Every apartment of $\mathcal{G}_k(V)$ is a $J(n,k)$-subset and, by Theorem 4.2, every $J(n,k)$-subset of $\mathcal{G}_k(V)$ is an apartment only in the case when $n = 2k \geq 4$. It was noted in Remark 4.7 that for every $k \in \{1,\ldots,n-1\}$ apartments of $\mathcal{G}_k(V)$ can be characterized in terms of independent subsets of projective spaces [Cooperstein, Kasikova and Shult (2005)].

All buildings of types $\mathsf{B}_n = \mathsf{C}_n$ and D_n can be obtained from polar spaces. A *polar space* is a pair $\Pi = (\mathcal{P}, \mathcal{L})$, where $\mathcal{P}$ is a non-empty set whose elements are called *points* and $\mathcal{L}$ is a set formed by proper subsets of $\mathcal{P}$ called *lines*. The lines satisfy some axioms (see [Buekenhout and Cohen (2013); Pankov (2010); Shult (2011); Ueberberg (2011)] for the precise definition of polar spaces). In contrast to projective spaces, there are pairs of non-collinear points, i.e. points which are not connected by a line. A subset $S \subset \mathcal{P}$ is a *singular subspace* of the polar space Π if any two distinct points of S are collinear and S contains the line joining them (the empty set and one-element subsets are singular subspaces). If there is a singular subspace containing more than one line then all maximal singular subspaces are projective spaces of the same dimension $n-1$ and the number n is called the *rank* of the polar space. Some of polar spaces are related to non-degenerate reflexive forms (alternating, symmetric, hermitian). In such a polar space, singular subspaces can be identified with totally isotropic subspaces of the corresponding reflexive form.

Every building of type C_n is the flag complex $\Delta(\Pi)$ of a rank n polar space Π. The vertex set of $\Delta(\Pi)$ consists of all non-empty singular subspaces of Π and the simplices are flags formed by them. The Grassmannians of this building are the *polar Grassmannians* $\mathcal{G}_k(\Pi)$, $k \in \{0,\ldots,n-1\}$. Every $\mathcal{G}_k(\Pi)$ consists of all k-dimensional singular subspaces of Π and the associated Grassmann graph is denoted by $\Gamma_k(\Pi)$. The graph $\Gamma_{n-1}(\Pi)$ is known as the *dual polar graph* and two maximal singular subspaces are adjacent vertices of this graph if their intersection is $(n-2)$-dimensional. In the case when $k \leq n-2$, two distinct k-dimensional singular subspaces are adjacent vertices of $\Gamma_k(\Pi)$ if they are contained in a certain $(k+1)$-dimensional singular subspace. A *frame* is a subset formed by $2n$ distinct points $p_1,\ldots,p_{2n}$ satisfying the following condition: for every $i \in \{1,\ldots,2n\}$ there is unique $\delta(i) \in \{1,\ldots,2n\}$ such that p_i and $p_{\delta(i)}$ are non-collinear. Any

k distinct mutually collinear points of a frame span a $(k - 1)$-dimensional singular subspace. Every apartment of the building $\Delta(\Pi)$ is defined by a certain frame, it consists of all flags formed by the singular subspaces spanned by subsets of this frame. The associated apartment of $\mathcal{G}_k(\Pi)$ is the set of all k-dimensional singular subspaces spanned by subsets of the frame (in the case when $k = 0$, this apartment is the frame). Apartments of $\mathcal{G}_{n-1}(\Pi)$ can be characterized as the images of isometric embeddings of the n-dimensional hypercube graph H_n in the dual polar graph [Pankov 2 (2011)]. Note that there exist non-isometric embeddings of H_n in $\Gamma_{n-1}(\Pi)$ [Cooperstein and Shult (1997)]. Now we take any apartment of $\mathcal{G}_k(\Pi)$ and denote by Γ_k the restriction of the graph $\Gamma_k(\Pi)$ to this apartment. Every subgraph of $\Gamma_0(\Pi)$ isomorphic to Γ_0 is a frame (this follows directly from the frame definition). In contrast to apartments of Grassmannians of vector spaces, apartments of $\mathcal{G}_k(\Pi)$ can be characterized as the images of isometric embeddings of Γ_k in $\Gamma_k(\Pi)$ for every k [Kwiatkowski and Pankov (2015)].

If S is an m-dimensional singular subspace of Π then the set of all $(m + 1)$-dimensional singular subspaces containing S admits the natural structure of a rank $n - m - 1$ polar space. This polar space is denoted by Π_S. The restriction of the dual polar graph $\Gamma_{n-1}(\Pi)$ to this set is the dual polar graph $\Gamma_{n-m-2}(\Pi_S)$. By [Pankov 2 (2011)], every isometric embedding of $\Gamma_{n-1}(\Pi)$ in $\Gamma_{n'-1}(\Pi')$, where Π' is a polar space of rank $n' \geq n$, is induced by a frame preserving embedding of Π in Π'_S, where S is an $(n' - n - 1)$-dimensional singular subspace of Π'. The same holds for all graphs associated to polar Grassmannians except one special case [Kwiatkowski and Pankov (2015)]. A frame preserving embedding, i.e. an embedding transferring frames to frames, is induced by a semilinear embedding if the corresponding polar spaces are embedable in projective spaces.

For every rank n polar space one of the following possibilities is realized:

- every $(n - 2)$-dimensional singular subspace is contained in at least three maximal singular subspaces,
- every $(n - 2)$-dimensional singular subspace is contained in precisely two maximal singular subspaces.

In the second case, we say that our polar space is of type D_n. If Π is such a polar space then $\mathcal{G}_{n-1}(\Pi)$ can be presented as the disjoint union of so-called *half-spin Grassmannians* $\mathcal{G}_+(\Pi)$ and $\mathcal{G}_-(\Pi)$. For any $S, U \in \mathcal{G}_{n-1}(\Pi)$ the distance $d(S, U)$ in the dual polar graph is equal to

$$n - 1 - \dim(S \cap U).$$

The half-spin Grassmannians are characterized by the following property:

$$d(S, U) \text{ is } \begin{cases} \text{even} & \text{if } S, U \in \mathcal{G}_+(\Pi) \text{ or } S, U \in \mathcal{G}_-(\Pi) \\ \text{odd} & \text{if } S \in \mathcal{G}_+(\Pi), \ U \in \mathcal{G}_-(\Pi). \end{cases}$$

For example, if Ω is a non-degenerate symmetric bilinear form defined on a $(2n)$-dimensional vector space (over a field whose characteristic is not equal to 2) and maximal totally isotropic subspaces are n-dimensional then the associated polar space is of type D_n and the corresponding half-spin Grassmannians are the orbits of the action of the orthogonal group $O_+(\Omega)$ on the set of maximal totally isotropic subspaces.

Every building of type D_n is defined by a certain polar space of the same type. The associated building Grassmannians are some of polar Grassmannians and the half-spin Grassmannians. Let $\delta \in \{+, -\}$. Denote by $\Gamma_\delta(\Pi)$ the Grassmann graph corresponding to $\mathcal{G}_\delta(\Pi)$. Two elements of $\mathcal{G}_\delta(\Pi)$ are adjacent vertices of this graph if the distance between them in the dual polar graph is equal to 2. All apartments of $\mathcal{G}_\delta(\Pi)$ are the intersections of $\mathcal{G}_\delta(\Pi)$ with apartments of $\mathcal{G}_{n-1}(\Pi)$. The restriction of the graph $\Gamma_\delta(\Pi)$ to any apartment is isomorphic to the n-dimensional half-cube graph $\frac{1}{2}H_n$. If n is even then apartments of $\mathcal{G}_\delta(\Pi)$ can be characterized as the images of isometric embeddings of $\frac{1}{2}H_n$ in $\Gamma_\delta(\Pi)$ [Pankov 2 (2014)]. In the case when n is odd, every vertex of $\frac{1}{2}H_n$ has a few opposite vertices and we conjecture that there exist isometric embeddings of $\frac{1}{2}H_n$ in $\Gamma_\delta(\Pi)$ whose images are not apartments of $\mathcal{G}_\delta(\Pi)$. Also, if n is even then every isometric embedding of $\Gamma_\delta(\Pi)$ in $\Gamma_\gamma(\Pi')$, where Π' is a polar space of type $\mathsf{D}_{n'}$ and $n' \geq n$, is induced by a frame preserving embedding of Π in Π'_S, where S is an $(n' - n - 1)$-dimensional singular subspace of Π'[Pankov 2 (2014)].

We do not consider Grassmannians associated to buildings of exceptional types F_4 and E_i, $i = 6, 7, 8$. Various characterizations of apartments in building Grassmannians can be found in [Cooperstein and Shult (1997); Cooperstein, Kasikova and Shult (2005); Kasikova (2007, 2009, 2013); Pankov 1 (2012)] and we refer [Cooperstein (2013)] for a survey.

Chapter 5

Characterization of isometric embeddings

As in Chapter 3, we consider isometric embeddings of $\Gamma_k(V)$ in $\Gamma_{k'}(V')$. Every such an embedding transfers apartments of $\mathcal{G}_k(V)$ to $J(n,k)$-subsets of $\mathcal{G}_{k'}(V')$ (which are not necessarily apartments). We show that this property characterizers isometric embeddings, in other words, every mapping of $\mathcal{G}_k(V)$ to $\mathcal{G}_{k'}(V')$ sending apartments to $J(n,k)$-subsets is an isometric embedding of $\Gamma_k(V)$ in $\Gamma_{k'}(V')$. This statement generalizes the description of mappings between Grassmannians which transfer apartments to apartments [Pankov (2010), Theorem 3.10].

The proof of the statement (Sections 5.2–5.5) is based on a characterization of the distance in terms of maximal intersections of $J(n,k)$-subsets of the same type. The second step of the proof is to show that all apartments go to $J(n,k)$-subsets of the same type.

In the next chapter this result will be reformulated in terms of exterior powers.

5.1 Main result, corollaries and remarks

Throughout this chapter we suppose that V and V' are left vector spaces over some division rings and their dimensions $\dim V = n$ and $\dim V' = n'$ are finite. Let also k and k' be integers satisfying

$$1 < k < n-1 \ \text{ and } \ 1 < k' < n'-1.$$

Then n and n' both are not less than 4.

If f is an isometric embedding of $\Gamma_k(V)$ in $\Gamma_{k'}(V')$ then the restriction of f to every apartment $\mathcal{A} \subset \mathcal{G}_k(V)$ is an isometric embedding of $J(n,k)$ in $\Gamma_{k'}(V')$ and $f(\mathcal{A})$ is a $J(n,k)$-subset. The main result of this chapter is the following.

Theorem 5.1 ([Pankov 1 (2014)]). *If a mapping of $\mathcal{G}_k(V)$ to $\mathcal{G}_{k'}(V')$ transfers every apartment to a $J(n, k)$-subset then it is an isometric embedding of $\Gamma_k(V)$ in $\Gamma_{k'}(V')$.*

The proof of Theorem 5.1 is complicated. It will be given in several steps (Sections 5.2–5.5).

If $n = n'$ and l is a strong semilinear embedding of V in V' then the mapping $(l)_k$ transfers apartments of $\mathcal{G}_k(V)$ to apartments of $\mathcal{G}_k(V')$. If $n = n' = 2k$ and s is a strong semilinear embedding of V in V'^* then the same holds for the mapping $(s)_k^*$. In the case when $n = n'$, all mappings of $\mathcal{G}_k(V)$ to $\mathcal{G}_k(V')$ sending apartments to apartments are known [Pankov (2010), Theorem 3.10]: they are induced by strong semilinear embeddings of V in V' or strong semilinear embeddings of V in V'^* and the second possibility can be realized only in the case when $n = 2k$.

Using Theorems 3.1 and 5.1 we prove the following more general statement.

Corollary 5.1. *Let $f : \mathcal{G}_k(V) \to \mathcal{G}_{k'}(V')$ be a mapping which transfers every apartment to a $J(n, k)$-subset. Suppose that there exists an apartment $\mathcal{A} \subset \mathcal{G}_k(V)$ such that*

$$f(\mathcal{A}) = \Phi_S^U(\mathcal{A}'),$$

where S and U are incident subspaces of V' and $\mathcal{A}'$ is an apartment in one of the Grassmannians of U/S (note that this holds for every apartment $\mathcal{A} \subset \mathcal{G}_k(V)$ if $n = 2k$). Then one of the following possibilities is realized:

(1) $S \in \mathcal{G}_{k'-k}(V')$, $U \in \mathcal{G}_{k'-k+n}(V')$ *and*

$$f = \Phi_S^U(l)_k,$$

    where $l : V \to U/S$ is a strong semilinear embedding;

(2) $S \in \mathcal{G}_{k'+k-n}(V')$, $U \in \mathcal{G}_{k'+k}(V')$ *and*

$$f = \Phi_S^U(s)_k^*,$$

    where $s : V \to (U/S)^*$ is a strong semilinear embedding.

Proof. By Theorem 5.1, f is an isometric embedding of $\Gamma_k(V)$ in $\Gamma_{k'}(V')$ and Theorem 3.1 shows that the statement holds for $n = 2k$.

Suppose that $k < n - k$. If f is an embedding of type (A) then Theorem 3.1 implies that

$$f = \Phi_{\tilde{S}}(l)_k,$$

where $\tilde{S} \in \mathcal{G}_{k'-k}(V')$ and $l : V \to V/\tilde{S}$ is a semilinear embedding. It is clear that S coincides with $\tilde{S}$ and $(l)_k$ transfers $\mathcal{A}$ to $\mathcal{A}'$. The latter guarantees that l is a strong semilinear embedding and we get the first possibility.

If our embedding is of type (B) then f^* is an isometric embedding of $\Gamma_k(V)$ in $\Gamma_{n'-k'}(V'^*)$ of type (A) and

$$f^*(\mathcal{A}) = \Phi_{U^0}^{S^0}(\mathcal{A}''),$$

where $\mathcal{A}''$ is an apartment in one of the Grassmannians of S^0/U^0. Thus

$$U^0 \in \mathcal{G}_{n'-k'-k}(V'^*), \quad S^0 \in \mathcal{G}_{n'-k'-k+n}(V'^*)$$

and the image of f^* is contained in $[U^0, S^0]_{n'-k'}$. This means that

$$S \in \mathcal{G}_{k'+k-n}(V'), \quad U \in \mathcal{G}_{k'+k}(V')$$

and the image of f is a subset of $[S, U]_{k'}$. Then

$$f = \Phi_S^U f',$$

where f' is an isometric embedding of $\Gamma_k(V)$ in $\Gamma_{n-k}(U/S)$. This embedding is of type (B) and Theorem 3.1 shows that $f' = (s)_k^*$ for a certain semilinear embedding $s : V \to (U/S)^*$. Since $f'(\mathcal{A}) = \mathcal{A}'$, the mapping $(s)_k$ transfers $\mathcal{A}$ to $\mathcal{A}'^0$ which implies that s is a strong semilinear embedding.

Consider the case when $k > n-k$. Since $\mathcal{A}^0$ is an apartment of $\mathcal{G}_{n-k}(V^*)$ and $f_*(\mathcal{A}^0) = f(\mathcal{A})$, we can apply the above arguments to f_* and use Remark 3.3. □

In the case when $n = n'$, there exist mappings of $\mathcal{G}_1(V)$ to $\mathcal{G}_1(V')$ which transfer bases of Π_V to bases of $\Pi_{V'}$ and cannot be induced by semilinear embeddings of V in V'.

Example 5.1 ([Huang and Kreuzer (1995)]). Consider the mapping

$$\alpha : \mathbb{R} \setminus \{0\} \to \mathcal{G}_1(\mathbb{R}^n)$$

which transfers every non-zero real number t to the 1-dimensional subspace containing the vector

$$(t, t^2, \ldots, t^n).$$

This mapping is injective and for any mutually distinct $t_1, \ldots, t_n \in \mathbb{R} \setminus \{0\}$ we have

$$\begin{vmatrix} t_1 & \cdots & t_1^n \\ \vdots & \ddots & \vdots \\ t_n & \cdots & t_n^n \end{vmatrix} \neq 0.$$

Therefore, every n-element subset of $\alpha(\mathbb{R})$ is a base of $\Pi_{\mathbb{R}^n}$. Since the sets $\mathbb{R} \setminus \{0\}$ and $\mathcal{G}_1(\mathbb{R}^n)$ are of the same infinite cardinality, there is a bijection of $\mathcal{G}_1(\mathbb{R}^n)$ to $\alpha(\mathbb{R})$. This is an injective transformation of $\mathcal{G}_1(\mathbb{R}^n)$ which sends bases of $\Pi_{\mathbb{R}^n}$ to bases of $\Pi_{\mathbb{R}^n}$. It is clear that this mapping cannot be induced by a linear transformation of $\mathbb{R}^n$.

Example 5.2 (H. Havlicek). Suppose that V is a vector space over a finite field and

$$|\mathcal{G}_1(V)| \leq n'.$$

Consider any bijection of $\mathcal{G}_1(V)$ to an independent subset of $\Pi_{V'}$. It transfers independent subsets of Π_V to independent subsets of $\Pi_{V'}$. Since any triple of collinear points goes to a triple of non-collinear points, this mapping cannot be induced by a semilinear embedding of V in V'.

Remark 5.1. The mappings considered in Examples 5.1 and 5.2 both are non-surjective. Suppose that f is a bijection of Π_V to $\Pi_{V'}$ transferring independent subsets to independent subsets. Then f^{-1} sends any triple of collinear points to a triple of collinear points. Hence it is a semicollineation of $\Pi_{V'}$ to Π_V and, by Corollary 2.3, f is a collineation of Π_V to $\Pi_{V'}$ if $n = n'$.

Problem 5.1. All mappings of polar and half-spin Grassmannians transferring apartments to apartments are described in [Pankov (2010), Section 4.9]. Are there analogues of Theorem 5.1 for such building Grassmannians? This is closely related to the existence of the metric characterization of apartments considered in Section 4.5. We can get an interesting analogue of Theorem 5.1 only in the case when this characterization does not hold. By our conjecture, the exceptional case is the half-spin Grassmannian $\mathcal{G}_\delta(\Pi)$, $\delta \in \{+, -\}$, where Π is a polar space of type D_n and n is odd.

5.2   Characterization of distance

Let $\mathcal{X} = \{P_1, \ldots, P_n\}$ be an m-independent subset of Π_W, where W is a finite-dimensional vector space over a division ring. Suppose that $m \geq 2k$ and consider the set $\mathcal{J} = J_k(\mathcal{X})$ consisting of all k-dimensional subspaces of W obtained from k-element subsets of $\mathcal{X}$ (Section 4.2). This is a $J(n, k)$-subset of $\mathcal{G}_k(W)$.

For every $i \in \{1, \ldots, n\}$ we denote by $\mathcal{J}(+i)$ and $\mathcal{J}(-i)$ the sets consisting of all elements of $\mathcal{J}$ which contain P_i and do not contain P_i, respectively.

Also, we write $\mathcal{J}(+i, +j)$ for the intersection of $\mathcal{J}(+i)$ and $\mathcal{J}(+j)$, i.e. the set of all elements of $\mathcal{J}$ containing both P_i and P_j. Every subset of type

$$\mathcal{J}(+i, +j) \cup \mathcal{J}(-i), \quad i \neq j$$

will be called *special*.

We say that $\mathcal{I}$ is an *inexact* subset of $\mathcal{J}$ if there exists a $J(n, k)$-subset $J_k(\mathcal{X}') \neq \mathcal{J}$ containing $\mathcal{I}$, in other words,

$$\mathcal{I} \subset \mathcal{J} \cap J_k(\mathcal{X}'),$$

where $\mathcal{X}'$ is an m'-independent subset of Π_W consisting of n elements, $m' \geq 2k$ and $\mathcal{X}' \neq \mathcal{X}$ (the latter guarantees that $J_k(\mathcal{X}') \neq \mathcal{J}$).

Lemma 5.1. *Every inexact subset is contained in a special subset.*

Proof. Let $\mathcal{I}$ be an inexact subset. For every $i \in \{1, \dots, n\}$ we denote by S_i the intersection of all elements of $\mathcal{I}$ containing P_i and we set $S_i = 0$ if $\mathcal{I}$ does not contain such elements. There is at least one integer i such that $S_i \neq P_i$ (otherwise, $\mathcal{I}$ is not inexact). Then

$$S_i = 0 \quad \text{or} \quad \dim S_i \geq 2.$$

If $S_i = 0$ then $\mathcal{I}$ is contained in $\mathcal{J}(-i)$. If $\dim S_i \geq 2$ then the inclusion

$$\mathcal{I} \subset \mathcal{J}(+i, +j) \cup \mathcal{J}(-i)$$

holds for any $j \neq i$ such that $P_j \subset S_i$. □

Lemma 5.2. *If $\mathcal{X}$ is an independent subset of Π_W then the class of maximal inexact subsets coincides with the class of special subsets.*

Proof. By Lemma 5.1, it is sufficient to show that every special subset is inexact. We take any 1-dimensional subspace P' contained in $P_i + P_j$ and different from P_i and P_j. Since $\mathcal{X}$ is independent,

$$\mathcal{X}' := (\mathcal{X} \setminus \{P_i\}) \cup \{P'\}$$

is independent. An easy verification shows that the intersection of $\mathcal{J}$ and $J_k(\mathcal{X}')$ coincides with the special subset $\mathcal{J}(+i, +j) \cup \mathcal{J}(-i)$. □

The following example shows that special subsets are not inexact in some cases.

Example 5.3. Suppose that $\mathcal{X} = \{P_1, \dots, P_5\}$ is a 4-simplex, $\mathcal{J} = J_2(\mathcal{X})$ and W is a vector space over $\mathbb{Z}_2$. The latter implies that every 2-dimensional subspace of W contains precisely three 1-dimensional subspaces. Let P' be the 1-dimensional subspace contained in $P_1 + P_2$ and different from P_1 and P_2. Since $\mathcal{X}$ is a 4-simplex, there exist non-zero vectors $x_1 \in P_1, \ldots, x_4 \in P_4$ such that

$$P_5 = \langle x_1 + \cdots + x_4 \rangle.$$

Then $P' = \langle x_1 + x_2 \rangle$ and

$$P_5 \subset P' + P_3 + P_4$$

which implies that the subset formed by P', P_3, P_4, P_5 is not independent. Hence the subset

$$(\mathcal{X} \setminus \{P_1\}) \cup \{P'\}$$

is not 4-independent and P_1 cannot be replaced by P' as in the proof of Lemma 5.2. Therefore, $\mathcal{J}(+1, +2) \cup \mathcal{J}(-1)$ is not inexact. The same arguments show that every special subset is not inexact.

The subsets $\mathcal{J}(+i, +j)$ and $\mathcal{J}(-i)$ are disjoint. This means that every special subset contains precisely

$$a(n, k) := |\mathcal{J}(+i, +j)| + |\mathcal{J}(-i)| = \binom{n-2}{k-2} + \binom{n-1}{k}$$

elements. Lemma 5.1 implies the following.

Lemma 5.3. *If an inexact subset consists of $a(n, k)$ elements then it is a special subset.*

A subset $\mathcal{C} \subset \mathcal{J}$ is said to be *complementary* if $\mathcal{J} \setminus \mathcal{C}$ is special, i.e.

$$\mathcal{J} \setminus \mathcal{C} = \mathcal{J}(+i, +j) \cup \mathcal{J}(-i)$$

for some distinct i, j. Then

$$\mathcal{C} = \mathcal{J}(+i) \cap \mathcal{J}(-j).$$

This complementary subset will be denoted by $\mathcal{J}(+i, -j)$.

Our proof of Theorem 5.1 is based on the following characterization of the distance between elements of $\mathcal{J}$ in terms of complementary subsets.

Lemma 5.4. *Let $S, U \in \mathcal{J}$. Then $d(S, U) = u$ if and only if there are precisely*

$$(k - u)(n - k - u)$$

distinct complementary subsets of $\mathcal{J}$ containing both S and U.

Proof. Since $2k \leq m \leq \dim W$, the distance between any two elements of $\mathcal{J}$ is not greater than k. The equality $d(S, U) = u$ implies that

$$\dim(S \cap U) = k - u \quad \text{and} \quad \dim(S + U) = k + u.$$

The complementary subset $\mathcal{J}(+i, -j)$ contains both S and U if and only if

$$P_i \subset S \cap U \quad \text{and} \quad P_j \not\subset S \cup U.$$

So, there are precisely $k - u$ possibilities for i. In particular, there exist no complementary subsets containing both S and U if $u = k$. Suppose that $u < k$. Then $k + u < 2k \leq m$. Since $\mathcal{X}$ is m-independent, the subspace $S + U$ contains precisely $k + u$ elements of $\mathcal{X}$ and we have precisely $n - k - u$ possibilities for j [1]. $\qquad\square$

5.3 Connectedness of the apartment graph

Let $x_1, \ldots, x_n$ be a base of V and let $\mathcal{A}$ be the associated apartment of $\mathcal{G}_k(V)$. Recall that for any distinct $i, j \in \{1, \ldots, n\}$ the subset formed by all elements of $\mathcal{A}$ which contain both x_i, x_j and all elements of $\mathcal{A}$ which do not contain x_i is called *special*. Every special subset of $\mathcal{A}$ is the intersection of $\mathcal{A}$ with an apartment different from $\mathcal{A}$ (similarly to Lemma 5.2).

Two apartments of $\mathcal{G}_k(V)$ are said to be *adjacent* if their intersection is a special subset. Consider the graph $A_k(V)$ whose vertices are apartments of $\mathcal{G}_k(V)$ and whose edges are pairs of adjacent apartments.

Proposition 5.1. *The graph* $A_k(V)$ *is connected.*

Proof. Let B and B' be bases of V. The associated apartments of $\mathcal{G}_k(V)$ will be denoted by $\mathcal{A}$ and $\mathcal{A}'$, respectively. Suppose that $\mathcal{A} \neq \mathcal{A}'$ and show that there is a path in $A_k(V)$ connecting these apartments.

First we consider the case when $|B \cap B'| = n - 1$. Let

$$B = \{x_1, \ldots, x_{n-1}, x_n\} \quad \text{and} \quad B' = \{x_1, \ldots, x_{n-1}, x'_n\}.$$

Since $\mathcal{A} \neq \mathcal{A}'$, the vector x'_n is a linear combination of x_n and some others $x_{i_1}, \ldots, x_{i_m}$. We suppose that

$$x'_n = ax_n + \sum_{i=1}^{m} a_i x_i,$$

---

[1] The m-dimensional subspaces of W spanned by subsets of $\mathcal{X}$ can contain more than m elements of $\mathcal{X}$.

where $m \leq n-1$ and all scalars $a, a_1, \ldots, a_m$ are non-zero. For every integer $j \leq m$ we define

$$x_n^j := ax_n + \sum_{i=1}^{j} a_i x_i$$

and denote by $\mathcal{A}_j$ the apartment of $\mathcal{G}_k(V)$ associated to the base

$$x_1, \ldots, x_{n-1}, x_n^j.$$

The vector x_n^1 is a linear combination of x_1 and x_n. The intersection $\mathcal{A} \cap \mathcal{A}_1$ is the special subset formed by all elements of $\mathcal{A}$ containing both x_1, x_n and all elements of $\mathcal{A}$ which do not contain x_n. So, the apartments $\mathcal{A}$ and $\mathcal{A}_1$ are adjacent. Similarly, the vector x_n^{j+1} is a linear combination of x_n^j and x_{j+1} and we show that $\mathcal{A}_j$ and $\mathcal{A}_{j+1}$ are adjacent for every $j \in \{1, \ldots, m-1\}$. Therefore,

$$\mathcal{A}, \mathcal{A}_1, \ldots, \mathcal{A}_m = \mathcal{A}'$$

is a path in $A_k(V)$.

Now we suppose that

$$|B \cap B'| = m < n-1$$

(possibly $m = 0$). Let

$$B \setminus B' = \{x_1, \ldots, x_{n-m}\} \quad \text{and} \quad x' \in B' \setminus B.$$

For every $i \in \{1, \ldots, n-m\}$ we consider the $(n-1)$-dimensional subspace

$$S_i := \langle B \setminus \{x_i\} \rangle.$$

Since the intersection of all S_i coincides with $\langle B \cap B' \rangle$ and x' does not belong to $\langle B \cap B' \rangle$, there is at least one S_i which does not contain x'. Then

$$B_1 := (B \setminus \{x_i\}) \cup \{x'\}$$

is a base of V. Denote by $\mathcal{A}_1$ the associated apartment of $\mathcal{G}_k(V)$. We have

$$|B \cap B_1| = n-1 \quad \text{and} \quad |B_1 \cap B'| = m+1.$$

The apartment $\mathcal{A}_1$ coincides with $\mathcal{A}$ if x' is a scalar multiple of x_i. Otherwise, there is a path in $A_k(V)$ connecting $\mathcal{A}$ and $\mathcal{A}_1$. Step by step, we construct a sequence of bases

$$B = B_0, B_1, \ldots, B_{n-m} = B'$$

of V such that

$$|B_{i-1} \cap B_i| = n-1$$

for every $i \in \{1, \ldots, n-m\}$. Let $\mathcal{A}_i$ be the apartment of $\mathcal{G}_k(V)$ defined by the base B_i. Then for every $i \in \{1, \ldots, n-m\}$ the apartments $\mathcal{A}_{i-1}$ and $\mathcal{A}_i$ are coincident or there is a path in $A_k(V)$ connecting them. This implies the existence of a path in $A_k(V)$ which connects $\mathcal{A} = \mathcal{A}_0$ and $\mathcal{A}' = \mathcal{A}_{n-m}$. $\square$

5.4 Intersections of $J(n, k)$-subsets of different types

In this subsection we assume that W is a vector space over a division ring and $\dim W = 2k \geq 4$. Let

$$\mathcal{X} = \{P_1, \ldots, P_n\} \quad \text{and} \quad \mathcal{Y} = \{P_1^*, \ldots, P_n^*\}, \quad n > 2k$$

be $(2k)$-independent subsets of Π_W and Π_{W^*}, respectively. Denote by U_i the annihilator of P_i^*. Each U_i is a $(2k - 1)$-dimensional subspace of W. Suppose that the following conditions hold:

- every U_i is the sum of some elements from $\mathcal{X}$,
- every P_i is the intersection of some U_j.

Since $\mathcal{X}$ is $(2k)$-independent, every U_i is spanned by a $(2k - 1)$-element subset $\mathcal{X}_i \subset \mathcal{X}$ and U_i does not contain any element from $\mathcal{X} \setminus \mathcal{X}_i$. Similarly, $\mathcal{Y}$ is a $(2k)$-independent subset and every P_i is contained in precisely $2k - 1$ distinct U_j whose intersection coincides with P_i.

We will investigate the intersection

$$\mathcal{Z} = J_k(\mathcal{X}) \cap J_k^*(\mathcal{Y}).$$

It is formed by all elements of $\mathcal{G}_k(W)$ which are spanned by k-element subsets of $\mathcal{X}$ and can be presented as the intersections of k distinct U_j.

We define

$$b(n, k) := \frac{n}{k} \binom{2k - 1}{k}$$

and prove the following.

Lemma 5.5. $|\mathcal{Z}| \leq b(n, k)$.

Proof. Let $\mathcal{Z}_i$ be the set of all elements of $\mathcal{Z}$ containing P_i. There are precisely $2k - 1$ distinct U_j containing P_i and every element of $\mathcal{Z}$ is the intersection of k distinct U_j. This means that

$$|\mathcal{Z}_i| \leq \binom{2k - 1}{k}.$$

Since every element of $\mathcal{Z}$ belongs to k distinct $\mathcal{Z}_i$, we have

$$|\mathcal{Z}| = \frac{|\mathcal{Z}_1| + \cdots + |\mathcal{Z}_n|}{k}$$

which implies the required inequality. $\qquad\square$

Recall that every special subset of $J_k(\mathcal{X})$ consists of

$$a(n,k) = \binom{n-2}{k-2} + \binom{n-1}{k}$$

elements.

Lemma 5.6. $a(n,k) > b(n,k)$ *except the case when* $n = 5$ *and* $k = 2$.

Proof. We have

$$a(n,2) = 1 + \frac{(n-1)(n-2)}{2} = \frac{n^2 - 3n + 4}{2} \quad \text{and} \quad b(n,2) = \frac{3n}{2}.$$

An easy verification shows that the equality $a(n,2) > b(n,2)$ does not hold only for $n = 5$.

From this moment we suppose that $k \geq 3$. Then

$$a(n,k) = \binom{n-2}{k-2} + \binom{n-1}{k} = \frac{(n-2)!}{(k-2)!(n-k)!} + \frac{(n-1)!}{k!(n-k-1)!}$$

$$= \frac{(n-2)\dots(n-k+1)k(k-1)}{k!} + \frac{(n-1)\dots(n-k)}{k!}$$

$$= [k(k-1) + (n-1)(n-k)]\frac{(n-2)\dots(n-k+1)}{k!}$$

and

$$b(n,k) = \frac{n}{k}\binom{2k-1}{k} = \frac{n(2k-1)!}{k!k!} = \frac{n(2k-1)\dots(k+1)}{k!}$$

$$= [n(k+1)]\frac{(2k-1)\dots(k+2)}{k!}.$$

Since $n \geq 2k+1$ and $k \geq 3$, we have

$$(n-1)(n-k) + k(k-1) = (n-1)(n-k) + (k+1)(k-1) - (k-1)$$

$$\geq (n-1)(k+1) + (k+1)(k-1) - (k-1) = (n+k-2)(k+1) - (k-1)$$

$$\geq (n+1)(k+1) - (k-1) = n(k+1) + 2 > n(k+1).$$

So,

$$k(k-1) + (n-1)(n-k) > n(k+1). \tag{5.1}$$

Also, $n \geq 2k+1$ implies that

$$n - 2 \geq 2k - 1, \dots, n - k + 1 \geq k + 2$$

and we have

$$(n-2)\dots(n-k+1) \geq (2k-1)\dots(k+2). \tag{5.2}$$

The required inequality

$$a(n,k) = [k(k-1) + (n-1)(n-k)]\frac{(n-2)\dots(n-k+1)}{k!}$$

$$> [n(k+1)]\frac{(2k-1)\dots(k+2)}{k!} = b(n,k)$$

follows from (5.1) and (5.2). □

Lemma 5.7. *If* $n = 5$ *and* $k = 2$ *then* $|\mathcal{Z}| \leq 5 < 7 = a(5,2)$.

Proof. Consider the case when $n = 5$ and $k = 2$. The subspaces $U_1, \ldots, U_5$ are 3-dimensional. Each P_i is contained in precisely 3 distinct U_j and every element of $\mathcal{Z}$ is the intersection of 2 distinct U_j. If every U_i contains not greater than 2 elements of $\mathcal{Z}$ then

$$|\mathcal{Z}| \leq \frac{2 \cdot 5}{2} = 5$$

(since every element of $\mathcal{Z}$ is contained in 2 distinct U_j) and we get the claim. Therefore, we need to show that every U_i contains not greater than 2 elements of $\mathcal{Z}$.

If $U_1 = P_1 + P_2 + P_3$ contains 3 elements of $\mathcal{Z}$ then these elements are

$$P_1 + P_2, \; P_1 + P_3, \; P_2 + P_3$$

and we suppose that they are the intersections of U_1 with U_2, U_3, U_4, respectively. It is easy to see that each P_i, $i \in \{1,2,3\}$ is contained in 3 distinct U_j, $j \in \{1,2,3,4\}$. The subspace U_5 contains one of P_i, $i \in \{1,2,3\}$ and such P_i is contained in 4 distinct U_j which is impossible. □

We join Lemmas 5.5–5.7 and get the following.

Lemma 5.8. $|\mathcal{Z}| < a(n,k)$.

5.5 Proof of Theorem 5.1

Let $f : \mathcal{G}_k(V) \to \mathcal{G}_{k'}(V')$ be a mapping which transfers every apartment of $\mathcal{G}_k(V)$ to a $J(n,k)$-subset.

Lemma 5.9. *The mapping* f *is injective.*

Proof. Let P and Q be distinct elements of $\mathcal{G}_k(V)$. By Lemma 1.1, there exists an apartment $\mathcal{A} \subset \mathcal{G}_k(V)$ containing them. Since $f(\mathcal{A})$ is a $J(n,k)$-subset, $\mathcal{A}$ and $f(\mathcal{A})$ have the same number of elements which implies that $f(P) \neq f(Q)$. □

The mapping f_* transfers apartments of $\mathcal{G}_{n-k}(V^*)$ to $J(n,k)$-subsets. Recall that the classes of $J(n,k)$-subsets and $J(n,n-k)$-subsets of $\mathcal{G}_{k'}(V')$ are coincident. Since f is an isometric embedding of $\Gamma_k(V)$ in $\Gamma_{k'}(V')$ if and only if f_* is an isometric embedding of $\Gamma_{n-k}(V^*)$ in $\Gamma_{k'}(V')$, it sufficient to prove Theorem 5.1 only in the case when $k \leq n - k$.

From this moment we suppose that $k \leq n-k$. The existence of isometric embeddings of $J(n,k)$ in $\Gamma_{k'}(V')$ implies that

$$k \leq \min\{k', n - k, n' - k'\}.$$

Lemma 5.10. *If $n = 2k$ then there exists $S \in \mathcal{G}_{k'-k}(V')$ such that the image of f is contained in $[S\rangle_{k'}$.*

Proof. Let $\mathcal{A}$ and $\mathcal{A}'$ be distinct apartments of $\mathcal{G}_k(V)$. Then $f(\mathcal{A})$ and $f(\mathcal{A}')$ are $J(n,k)$-subsets. Since $n = 2k$, Theorem 4.2 implies that

$$f(\mathcal{A}) = \Phi_S(J_k(\mathcal{X})) \quad \text{and} \quad f(\mathcal{A}') = \Phi_{S'}(J_k(\mathcal{X}')),$$

where $S, S' \in \mathcal{G}_{k'-k}(V')$ and $\mathcal{X}, \mathcal{X}'$ are independent $(2k)$-element subsets of $\Pi_{V'/S}$ and $\Pi_{V'/S'}$, respectively. We need to show that $S = S'$.

By Proposition 5.1, it is sufficient to consider the case when the apartments $\mathcal{A}$ and $\mathcal{A}'$ are adjacent. Then

$$|f(\mathcal{A}) \cap f(\mathcal{A}')| = |\mathcal{A} \cap \mathcal{A}'| = a(2k, k)$$

and

$$\mathcal{Z} := (\Phi_S)^{-1}(f(\mathcal{A}) \cap f(\mathcal{A}'))$$

is a subset of $J_k(\mathcal{X})$ consisting of $a(2k,k)$ elements.

Denote by T the intersection of all elements from $\mathcal{Z}$. Since $S + S'$ is contained in all elements of $f(\mathcal{A}) \cap f(\mathcal{A}')$, every element of $\mathcal{Z}$ contains $(S + S')/S$ and we have

$$(S + S')/S \subset T.$$

Suppose that $S \neq S'$. Then

$$t = \dim T \geq \dim((S + S')/S) \geq 1$$

and there are precisely $\binom{2k-t}{k-t}$ elements of $J_k(\mathcal{X})$ containing T. This means that

$$|\mathcal{Z}| \leq \binom{2k - t}{k - t}.$$

Then

$$|\mathcal{Z}| \leq \binom{2k - 1}{k - 1} = \binom{2k - 1}{k} < \binom{2k - 1}{k} + \binom{2k - 2}{k - 2} = a(2k, k)$$

which is impossible. Thus $S = S'$. □

Lemma 5.11. *Suppose that $k < n - k$ and f transfers a certain apartment $\mathcal{A} \subset \mathcal{G}_k(V)$ to a $J(n,k)$-subset of type* (A). *Then the images of all apartments of $\mathcal{G}_k(V)$ are $J(n,k)$-subsets of type* (A) *and there exists $S \in \mathcal{G}_{k'-k}(V')$ such that the image of f is contained in $[S\rangle_{k'}$.*

Proof. By our hypothesis,

$$f(\mathcal{A}) = \Phi_S(J_k(\mathcal{X})),$$

where $S \in \mathcal{G}_{k'-k}(V')$ and $\mathcal{X}$ is a $(2k)$-independent subset of $\Pi_{V'/S}$ consisting of n elements. Let $S_1, \ldots, S_n$ be the $(k'-k+1)$-dimensional subspaces of V' corresponding to the elements of $\mathcal{X}$. Then the elements of $\mathcal{X}$ are

$$P_1 = S_1/S, \ldots, P_n = S_n/S$$

and every element of $f(\mathcal{A})$ is the sum of k distinct S_j.

Let $\mathcal{A}'$ be an apartment of $\mathcal{G}_k(V)$ distinct from $\mathcal{A}$. We need to show that $f(\mathcal{A}')$ is a $J(n,k)$-subset of type (A) and it is contained in $[S\rangle_{k'}$. By Proposition 5.1, it is sufficient to consider the case when $\mathcal{A}$ and $\mathcal{A}'$ are adjacent. As in the proof of the previous lemma,

$$\mathcal{Z} := (\Phi_S)^{-1}(f(\mathcal{A}) \cap f(\mathcal{A}'))$$

is a subset of $J_k(\mathcal{X})$ consisting of $a(n,k)$ elements. There are the following possibilities:

(1) $\mathcal{Z}$ is contained in a special subset of $J_k(\mathcal{X})$,
(2) there is no special subset of $J_k(\mathcal{X})$ containing $\mathcal{Z}$.

Case (1). Every special subset of $J_k(\mathcal{X})$ consists of $a(n,k) = |\mathcal{Z}|$ elements. This implies that $\mathcal{Z}$ is a special subset of $J_k(\mathcal{X})$. Suppose that

$$\mathcal{Z} = \mathcal{J}(+i, +j) \cup \mathcal{J}(-i)$$

(see Section 5.2 for the notation). We take any $(k-1)$-dimensional subspace $T \subset V'/S$ which is the sum of P_j and other $k-2$ elements from $\mathcal{X}$. Then

$$\mathcal{S} := J_k(\mathcal{X}) \cap [T\rangle_k$$

is a star of $J_k(\mathcal{X})$ contained in $\mathcal{Z}$. Indeed, an element of $\mathcal{S}$ containing P_i belongs to $\mathcal{J}(+i, +j)$ and an element of $\mathcal{S}$ belongs to $\mathcal{J}(-i)$ if it does not contain P_i.

Consider $\Phi_S(\mathcal{S})$. This is a star of $f(\mathcal{A})$. Since $f(\mathcal{A})$ is a $J(n,k)$-subset of type (A), this star consists of $n - k + 1$ vertices (Lemma 4.2). Also, $\Phi_S(\mathcal{S})$ is contained in a star of $f(\mathcal{A}')$. The inequality $n - k + 1 > k + 1$ and Lemma 4.2 guarantee that $f(\mathcal{A}')$ is a $J(n,k)$-subset of type (A).

We take $P, Q \in \mathcal{Z}$ such that $P \cap Q = 0$. The intersection of $\Phi_S(P)$ and $\Phi_S(Q)$ coincides with S. On the other hand, $\Phi_S(P)$ and $\Phi_S(Q)$ both belong to $f(\mathcal{A}')$. Since $f(\mathcal{A}')$ is a $J(n, k)$-subset of type (A), the associated $(k' - k)$-dimensional subspace of V' coincides with S and $f(\mathcal{A}')$ is contained in $[S\rangle_{k'}$.

Case (2). For every integer $i \in \{1, \ldots, n\}$ the intersection of all elements of $\mathcal{Z}$ containing P_i coincides with P_i (otherwise, as in the proof of Lemma 5.1 we show that $\mathcal{Z}$ is contained in a special subset of $J_k(\mathcal{X})$ which contradicts our assumption). Then the intersection of all elements of

$$\Phi_S(\mathcal{Z}) = f(\mathcal{A}) \cap f(\mathcal{A}')$$

containing S_i coincides with S_i. This means that the intersection of all elements of $f(\mathcal{A}) \cap f(\mathcal{A}')$ is S.

If $f(\mathcal{A}')$ is a $J(n, k)$-subset of type (A) then the associated $(k' - k)$-dimensional subspace of V' coincides with S. Hence $f(\mathcal{A}')$ is contained in $[S\rangle_{k'}$. Then $\mathcal{Z}$ is an inexact subset of $J_k(\mathcal{X})$. By Lemma 5.3, $\mathcal{Z}$ is a special subset of $J_k(\mathcal{X})$ which contradicts our assumption.

Therefore, $f(\mathcal{A}')$ is a $J(n, k)$-subset of type (B). Then

$$f(\mathcal{A}') = \Phi^U\left(J_k^*(\mathcal{Y})\right),$$

where $U \in \mathcal{G}_{k'+k}(V')$ and $\mathcal{Y}$ is a $(2k)$-independent subset of Π_{U^*} consisting of n elements $P_1^*, \ldots, P_n^*$. Denote by U_i the annihilator of P_i^* in U. Then every element of $f(\mathcal{A}')$ is the intersection of k distinct U_j.

Consider the set

$$(\Phi^U)^{-1}(f(\mathcal{A}) \cap f(\mathcal{A}')) \tag{5.3}$$

which is contained in $J_k^*(\mathcal{Y})$. Denote by $\mathcal{C}$ the subset of $J_k(\mathcal{Y})$ formed by the annihilators of all elements from (5.3). The set (5.3) consists of $a(n, k)$ elements and the same holds for $\mathcal{C}$.

If $\mathcal{C}$ is contained in a certain special subset of $J_k(\mathcal{Y})$ then it coincides with this special subset. As in the case (1), we take a star $\mathcal{S} \subset J_k(\mathcal{Y})$ contained in $\mathcal{C}$. Then $\mathcal{S}^0$ is a top of $J_k^*(\mathcal{Y})$ contained in (5.3). Hence $\Phi^U(\mathcal{S}^0)$ is a top of $f(\mathcal{A}')$ contained in $f(\mathcal{A}) \cap f(\mathcal{A}')$. This top consists of $n - k + 1$ vertices and $n - k + 1 > k + 1$. Also, $\Phi^U(\mathcal{S}^0)$ is contained in a top of $f(\mathcal{A})$. By Lemma 4.2, this contradicts the fact that $f(\mathcal{A})$ and $f(\mathcal{A}')$ are $J(n, k)$-subsets of different types.

Thus there is no special subset of $J_k(\mathcal{Y})$ containing $\mathcal{C}$. This means that for every $i \in \{1, \ldots, n\}$ the intersection of all elements of $\mathcal{C}$ containing P_i^* coincides with P_i^*. Then U_i is the sum of the annihilators (in U) of these elements, i.e. U_i is the sum of some elements from $f(\mathcal{A}) \cap f(\mathcal{A}')$. Since every element of $f(\mathcal{A})$ is the sum of k distinct S_j, we get the following property:

(*) every U_i is the sum of some S_j.

This means that every U_i contains S (since S is contained in all S_i) and $f(\mathcal{A}')$ is a subset of $[S\rangle_{k'}$ (every element of $f(\mathcal{A}')$ is the intersection of k distinct U_j).

The intersection of all elements of $f(\mathcal{A}) \cap f(\mathcal{A}')$ containing S_i coincides with S_i and every element of $f(\mathcal{A}')$ is the intersection of k distinct U_j. This implies the following property:

(**) every S_i is the intersection of some U_j.

Then every S_i is contained in U and $f(\mathcal{A})$ is a subset of $\langle U]_{k'}$ (since every element of $f(\mathcal{A})$ is the sum of k distinct S_j).

So, $f(\mathcal{A})$ and $f(\mathcal{A}')$ both are contained in $[S, U]_{k'}$. Consider the $(2k)$-dimensional vector space $W := U/S$. We have

$$f(\mathcal{A}) = \Phi_S^U(J_k(\mathcal{X})) \quad \text{and} \quad f(\mathcal{A}') = \Phi_S^U(J_k^*(\mathcal{Y}')),$$

where $\mathcal{Y}'$ is the subset of Π_{W^*} corresponding to $\mathcal{Y}$. The annihilators of the elements of $\mathcal{Y}'$ are U_i/S, $i \in \{1, \ldots, n\}$. The properties (*) and (**) guarantee that $\mathcal{X}$ and $\mathcal{Y}'$ satisfy the conditions of Section 5.4:

- every U_i/S is the sum of some elements from $\mathcal{X}$,
- every P_i is the intersection of some U_j/S.

Then Lemma 5.8 states that

$$\mathcal{Z} = J_k(\mathcal{X}) \cap J_k^*(\mathcal{Y}')$$

contains less than $a(n, k)$ elements. This contradiction shows that the case (2) is impossible. □

It follows from Lemma 5.11 that the images of all apartments of $\mathcal{G}_k(V)$ are $J(n, k)$-subsets of the same type if $k < n - k$. Suppose that one of the following possibilities is realized:

- $n = 2k$,
- $k < n - k$ and the images of all apartments of $\mathcal{G}_k(V)$ are $J(n, k)$-subsets of type (A).

Lemmas 5.10 and 5.11 imply the existence of $S \in \mathcal{G}_{k'-k}(V')$ such that the image of f is contained in $[S\rangle_{k'}$. There exists a mapping

$$g : \mathcal{G}_k(V) \to \mathcal{G}_k(V'/S)$$

such that $f = \Phi_S g$. The mapping g transfers every apartment of $\mathcal{G}_k(V)$ to a certain $J_k(\mathcal{X})$, where $\mathcal{X}$ is a $(2k)$-independent subset of $\Pi_{V'/S}$ consisting of n elements.

We use results of Section 5.2 to prove the following.

Lemma 5.12. *The mapping g is an isometric embedding of $\Gamma_k(V)$ in $\Gamma_k(V'/S)$.*

Proof. Let $P, Q \in \mathcal{G}_k(V)$. By Lemma 1.1, there exists an apartment $\mathcal{A}$ of $\mathcal{G}_k(V)$ containing P and Q. Let $\mathcal{S}$ be a special subset of $\mathcal{A}$. Lemma 5.2 shows that $\mathcal{S} = \mathcal{A} \cap \mathcal{A}'$, where $\mathcal{A}'$ is an apartment of $\mathcal{G}_k(V)$ adjacent to $\mathcal{A}$. Since $g(\mathcal{A})$ and $g(\mathcal{A}')$ are $J(n,k)$-subsets of $\mathcal{G}_k(V'/S)$,

$$g(\mathcal{S}) = g(\mathcal{A}) \cap g(\mathcal{A}')$$

is an inexact subset of $g(\mathcal{A})$. It consists of $a(n,k)$ elements and Lemma 5.3 implies that $g(\mathcal{S})$ is a special subset of $g(\mathcal{A})$. Since $\mathcal{A}$ and $g(\mathcal{A})$ have the same number of special subsets, a subset of $\mathcal{A}$ is special if and only if its image is a special subset of $g(\mathcal{A})$. This means that $\mathcal{C} \subset \mathcal{A}$ is a complementary subset if and only if $g(\mathcal{C})$ is a complementary subset of $g(\mathcal{A})$. Then Lemma 5.4 guarantees that

$$d(P, Q) = d(g(P), g(Q))$$

and we get the claim. $\qquad\square$

Since Φ_S is an isometric embedding of $\Gamma_k(V'/S)$ in $\Gamma_{k'}(V')$, Lemma 5.12 shows that $f = \Phi_S g$ is an isometric embedding of $\Gamma_k(V)$ in $\Gamma_{k'}(V')$.

Now we suppose that $k < n - k$ and the images of all apartments of $\mathcal{G}_k(V)$ are $J(n,k)$-subsets of type (B). Then f^* transfers every apartment to a $J(n,k)$-subset of type (A). Hence f^* is an isometric embedding of $\Gamma_k(V)$ in $\Gamma_{n'-k'}(V'^*)$ which means that f is an isometric embedding of $\Gamma_k(V)$ in $\Gamma_{k'}(V')$.

Chapter 6

Semilinear mappings of exterior powers

In this chapter some of the previous results will be reformulated in terms of semilinear mappings of exterior powers. For a vector space V over a field we consider the exterior product $\wedge^k(V)$. Every semilinear transformation of V induces a semilinear transformation of $\wedge^k(V)$ and there exist semilinear transformations of $\wedge^k(V)$ which cannot be obtained from semilinear transformations of V. The Grassmannian $\mathcal{G}_k(V)$ can be considered as a subset of the projective space $\Pi_{\wedge^k(V)}$. Following [Westwick (1964)] we use Chow's theorem to determining all semilinear automorphisms of $\wedge^k(V)$ preserving $\mathcal{G}_k(V)$. As an application, we describe the automorphism groups of Grassmann codes [Ghorpade and Kaipa (2013)]. Also, we apply Theorems 3.1 and 5.1 to similinear mappings of exterior powers which are not necessarily semlinear isomorphisms.

6.1 Exterior powers

Lct V be au n-dimensional vector space over a field F. Consider the Cartesian k-product

$$V^k = \underbrace{V \times \cdots \times V}_{k}.$$

A mapping $f : V^k \to W$, where W is a vector space over F, is called *multilinear* if it is linear in each of the variables separately, in other words, we have

$$f(x_1, \ldots, ax_i + by_i, \ldots, x_k) = af(x_1, \ldots, x_i, \ldots, x_k) + bf(x_1, \ldots, y_i, \ldots, x_k)$$

for every $i \in \{1, \ldots, k\}$; in the case when $k = 2$, we say that f is *bilinear*.

For every natural $k \geq 2$ we denote by $U^k(V)$ the vector space over F consisting of all formal linear combinations of elements from V^k. Then the elements of V^k form a base of $U^k(V)$. Note that this vector space is finite-dimensional only in the case when F is a finite field. Let $W^k(V)$ be the subspace of $U^k(V)$ spanned by all vectors of type

$$a(x_1, \ldots, x_k) - (x_1, \ldots, x_{i-1}, ax_i, x_{i+1}, \ldots, x_n)$$

and

$$(x_1, \ldots, x_{i-1}, x_i + y_i, x_{i+1}, \ldots, x_n) - (x_1, \ldots, x_i, \ldots, x_n) - (x_1, \ldots, y_i, \ldots, x_n).$$

The quotient vector space

$$\otimes^k(V) := U^k(V)/W^k(V)$$

is known as the *tensor k-product* of V. If $x_1, \ldots, x_k$ are vectors of V (not necessarily distinct) then the element of $\otimes^k(V)$ containing $(x_1, \ldots, x_k) \in V^k$ will be denote by $x_1 \otimes \cdots \otimes x_k$. The mapping

$$\otimes : V^k \to \otimes^k(V)$$

$$(x_1, \ldots, x_k) \to x_1 \otimes \cdots \otimes x_k$$

is multilinear. If $e_1, \ldots, e_n$ is a base of V then all vectors of type

$$e_{i_1} \otimes \cdots \otimes e_{i_k} \tag{6.1}$$

form a base of $\otimes^k(V)$. It must be pointed out that $i_1, \ldots, i_k$ are not necessarily distinct and the order of indices is important. So, the vector space $\otimes^k(V)$ is finite-dimensional.

Let $S^k(V)$ be the subspace of $\otimes^k(V)$ spanned by all vectors $x_1 \otimes \cdots \otimes x_k$ such that $x_i = x_j$ for some distinct $i, j \in \{1, \ldots, k\}$. The quotient vector space

$$\wedge^k(V) := \otimes^k(V)/S^k(V)$$

is called the *exterior k-product* of V. The composition of the multilinear mapping $\otimes : V^k \to \otimes^k(V)$ and the quotient projection of $\otimes^k(V)$ to $\wedge^k(V)$ is the multilinear mapping

$$\wedge : V^k \to \wedge^k(V)$$

and we write $x_1 \wedge \cdots \wedge x_k$ for the image of $(x_1, \ldots, x_k) \in V^k$.

Lemma 6.1. *For any vectors $x_1, \ldots, x_k \in V$ and every permutation σ on $\{1, \ldots, k\}$ we have*

$$x_{\sigma(1)} \wedge \cdots \wedge x_{\sigma(k)} = \operatorname{sgn}(\sigma) \, x_1 \wedge \cdots \wedge x_k,$$

where $\operatorname{sgn}(\sigma) = 1$ if the permutation is even and $\operatorname{sgn}(\sigma) = -1$ if it is odd.

Proof. We have

$$0 = (x_1 + x_2) \wedge (x_1 + x_2) \wedge x_3 \wedge \cdots \wedge x_k = x_1 \wedge x_2 \wedge \cdots \wedge x_k + x_2 \wedge x_1 \wedge \cdots \wedge x_k.$$

Similarly, we establish that

$$x_{\sigma(1)} \wedge \cdots \wedge x_{\sigma(k)} = -x_1 \wedge \cdots \wedge x_k$$

for every transposition σ. Every permutation on $\{1, \ldots, k\}$ can be presented as the composition of some transpositions. The permutation type (even or odd) depends on the number of such transpositions. □

If $e_1, \ldots, e_n$ is a base of V then all vectors of type (6.1) form a base of $\otimes^k(V)$ and Lemma 6.1 implies that all vectors of type

$$e_{i_1} \wedge \cdots \wedge e_{i_k}, \quad i_1 < \cdots < i_k$$

form a base of $\wedge^k(V)$. Therefore,

$$\dim(\wedge^k(V)) = \binom{n}{k}.$$

Also, $\wedge^k(V) = 0$ if $k > n$ and we will alway suppose that $k \le n$.

The vector space $\wedge^k(V)$ is defined for every $k \ge 2$ and we set $\wedge^1(V) = V$. Elements of the vector space $\wedge^k(V)$ are called *k-vectors* if $k \ge 2$.

Lemma 6.2. *The condition*

$$x_1 \wedge \cdots \wedge x_k = 0 \tag{6.2}$$

is equivalent to the fact that the vectors $x_1, \ldots, x_k$ are linearly dependent.

Proof. An easy verification shows that (6.2) holds if $x_1, \ldots, x_k$ are linearly dependent. If these vectors are linearly independent then we take any base of V containing them. This base induces a base of $\wedge^k(V)$ and the equality (6.2) is impossible. □

A non-zero k-vector is said to be *decomposable* if it can be written as $x_1 \wedge \cdots \wedge x_k$. Since the vector space $\wedge^n(V)$ is 1-dimensional, every n-vector is decomposable. Also, all $(n-1)$-vectors are decomposable and there exist non-decomposable k-vectors if $1 < k < n-1$.

Suppose that V' is a vector space over a field and $l : V \to V'$ is a σ-linear mapping. The correspondence

$$(x_1, \ldots, x_k) \to (l(x_1), \ldots, l(x_k)), \quad (x_1, \ldots, x_k) \in V^k$$

can be uniquely extended to a σ-linear mapping of $U^k(V)$ to $U^k(V')$. This mapping transfers $W^k(V)$ to a subset of $W^k(V')$ and, by Lemma 1.3, it induces a σ-linear mapping of $\otimes^k(V)$ to $\otimes^k(V')$ which will be denoted by $\otimes^k(l)$. Similarly, $\otimes^k(l)$ sends $S^k(V)$ to a subset of $S^k(V')$ and induces a σ-linear mapping of $\wedge^k(V)$ to $\wedge^k(V')$. We denote this mapping by $\wedge^k(l)$. It is clear that $\otimes^k(l)$ and $\wedge^k(l)$ send

$$x_1 \otimes \cdots \otimes x_k \quad \text{and} \quad x_1 \wedge \cdots \wedge x_k$$

to

$$l(x_1) \otimes \cdots \otimes l(x_k) \quad \text{and} \quad l(x_1) \wedge \cdots \wedge l(x_k),$$

respectively. Therefore, if l is a semilinear isomorphism or a strong semilinear embedding then the same holds for $\otimes^k(l)$ and $\wedge^k(l)$. Also, $\wedge^k(l)$ transfers decomposable k-vectors to decomposable k-vectors.

Example 6.1. Let W be an $(n-1)$-dimensional vector space over a field E. Suppose that σ is a homomorphism of F to E such that

$$[E : \sigma(F)] \geq n.$$

By Example 1.17, there is a σ-linear $(n-1)$-embedding $l : V \to W$. If $e_1, \ldots, e_n$ is a base of V then

$$u_1 := l(e_1), \ldots, u_{n-1} := l(e_{n-1})$$

form a base of W and $u_n := l(e_n)$ is a linear combination of $u_1, \ldots, u_{n-1}$, where every scalar is non-zero. For every $k < n$ the semilinear mapping $\wedge^k(l)$ transfers the base of $\wedge^k(V)$ defined by $e_1, \ldots, e_n$ to the set of all vectors of type

$$u_{i_1} \wedge \cdots \wedge u_{i_k}, \quad i_1 < \cdots < i_k.$$

The k-vector $u_1 \wedge \cdots \wedge u_{k-1} \wedge u_n$ is a linear combination of

$$u_1 \wedge \cdots \wedge u_{k-1} \wedge u_j, \quad j = k, \ldots, n-1.$$

Therefore, $\wedge^k(l)$ is not a semilinear $(n-k+1)$-embedding.

Since V is a vector space over a field, linear isomorphisms of V to V^* exist. Every such a linear isomorphism induces a linear isomorphism of $\wedge^k(V)$ to $\wedge^k(V^*)$. Now we show that every n-vector of V defines a linear isomorphism between $\wedge^k(V^*)$ and $\wedge^{n-k}(V)$. We start from the following.

Lemma 6.3. *Suppose that W is a vector space over F and $f : V^k \to W$ is a multilinear mapping satisfying the following condition*

(*) $f(x_1, \ldots, x_k) = 0$ *if $x_i = x_j$ for some distinct $i, j \in \{1, \ldots, k\}$.*

Then there is a unique linear mapping $l : \wedge^k(V) \to W$ such that

$$l(x_1 \wedge \cdots \wedge x_k) = f(x_1, \ldots, x_k)$$

for any vectors $x_1, \ldots, x_k \in V$.

Proof. There is the unique extension of f to a linear mapping

$$l_U : U^k(V) \to W.$$

Since f is multilinear, $l_U(W^k(V)) = 0$ and l_U induces a linear mapping

$$l_T : \otimes^k(V) \to W.$$

We have

$$l_T(x_1 \otimes \cdots \otimes x_k) = f(x_1, \ldots, x_k)$$

for any vectors $x_1, \ldots, x_k \in V$ and (*) implies that $l_T(S^k(V)) = 0$. Then l_T induces a linear mapping $l : \wedge^k(V) \to W$ satisfying the required condition.

$\square$

Let $x^* \in V^*$. Consider the multilinear mapping

$$f_{x^*} : V^k \to \wedge^{k-1}(V)$$

defined as follows

$$f_{x^*}(x_1, \ldots, x_k) = (x^* \cdot x_1)x_2 \wedge x_3 \wedge \cdots \wedge x_k - (x^* \cdot x_2)x_1 \wedge x_3 \wedge \cdots \wedge x_k$$

$$+ \cdots + (-1)^{k-1}(x^* \cdot x_k)x_1 \wedge x_2 \wedge \cdots \wedge x_{k-1}.$$

We check that this mapping satisfies the condition (*) from Lemma 6.3 (an exercise for the readers) and establish the existence of the linear mapping

$$l_{x^*} : \wedge^k(V) \to \wedge^{k-1}(V)$$

such that

$$l_{x^*}(x_1 \wedge \cdots \wedge x_k) = (x^* \cdot x_1)x_2 \wedge x_3 \wedge \cdots \wedge x_k - (x^* \cdot x_2)x_1 \wedge x_3 \wedge \cdots \wedge x_k$$

$$+ \cdots + (-1)^{k-1}(x^* \cdot x_k)x_1 \wedge x_2 \wedge \cdots \wedge x_{k-1}$$

for all vectors $x_1, \ldots, x_k \in V$.

Let $x_1^*, \ldots, x_m^* \in V^*$ and $m < k$. The composition $l_{x_m^*} \ldots l_{x_1^*}$ is a linear mapping of $\wedge^k(V)$ to $\wedge^{k-m}(V)$. We consider the multilinear mapping

$$(x_1^*, \ldots, x_m^*) \to l_{x_m^*} \ldots l_{x_1^*}$$

and show that it satisfies the condition (*) from Lemma 6.3. Hint: we determine every

$$l_{e_{j_m}^*} \dots l_{e_{j_1}^*} (e_{i_1} \wedge \cdots \wedge e_{i_k}),$$

where $e_1, \dots, e_n$ is a base of V, $e_1^*, \dots, e_n^*$ is the dual base of V^* and $e_{j_1}^*, \dots, e_{j_m}^*$ are not necessarily mutually distinct; if $x_i^* = x_j^*$ in $l_{x_m^*} \dots l_{x_1^*}$ then we suppose that $e_1^* = x_i^* = x_j^*$ and present the remaining $x_1^*, \dots, x_m^*$ as linear combinations of $e_1^*, \dots, e_n^*$. So, our multilinear mapping induces a linear mapping of $\wedge^m(V^*)$ to the vector space formed by linear mappings of $\wedge^k(V)$ to $\wedge^{k-m}(V)$. In other words, every $v^* \in \wedge^m(V^*)$ defines a linear mapping

$$l_{v^*} : \wedge^k(V) \to \wedge^{k-m}(V)$$

such that

$$l_{v^*} = l_{x_m^*} \dots l_{x_1^*} \text{ if } v^* = x_1^* \wedge \cdots \wedge x_m^*.$$

Therefore, we get the bilinear mapping

$$\wedge^m(V^*) \times \wedge^k(V) \to \wedge^{k-m}(V)$$

$$(v^*, v) \to l_{v^*}(v).$$

Then every n-vector $w \in \wedge^n(V)$ defines the linear mapping

$$h_w : \wedge^k(V^*) \to \wedge^{n-k}(V)$$

$$v^* \to l_{v^*}(w)$$

for every $k \in \{2, \dots, n-1\}$. The vector space $\wedge^n(V)$ is 1-dimensional and for any other n-vector $w' \in \wedge^n(V)$ the linear mapping $h_{w'}$ is a scalar multiple of h_w. Similarly, every n-vector of V^* defines a linear mapping of $\wedge^k(V)$ to $\wedge^{n-k}(V^*)$.

Proposition 6.1. *For every n-vector $w \in \wedge^n(V)$ the mapping h_w is a linear isomorphism of $\wedge^k(V^*)$ to $\wedge^{n-k}(V)$. It transfers decomposable k-vectors of V^* to decomposable $(n-k)$-vectors of V. The inverse mapping is defined by a certain n-vector of V^*.*

Proof. Let $w \in \wedge^n(V)$. There is a base $e_1, \dots, e_n$ of V such that

$$w = e_1 \wedge \cdots \wedge e_n.$$

Suppose that $e_1^*, \dots, e_n^*$ is the dual base of V^* and $e_i^* \cdot e_j = \delta_{ij}$. A direct verification shows that h_w transfers

$$e_{i_1}^* \wedge \cdots \wedge e_{i_k}^* \quad \text{to} \quad \pm e_{j_1} \wedge \cdots \wedge e_{j_{n-k}},$$

where $i_1, \ldots, i_k, j_1, \ldots, j_{n-k}$ is a permutation on $\{1, \ldots, n\}$ and the sign is the sign of this permutation. This implies that h_w is a linear isomorphism and the inverse mapping is h_{w^*}, where

$$w^* = \pm e_1^* \wedge \cdots \wedge e_n^*$$

and the sign is the sign of the permutation transferring

$$i_1, \ldots, i_k, j_1, \ldots, j_{n-k} \quad \text{to} \quad j_1, \ldots, j_{n-k}, i_1, \ldots, i_k.$$

Consider a decomposable k-vector $x_1^* \wedge \cdots \wedge x_k^*$. The vectors $x_1^*, \ldots, x_k^*$ are linearly independent and we extend them to a base $x_1^*, \ldots, x_n^*$. If $x_1, \ldots, x_n$ is the dual base in V then

$$h_{w'}, \quad w' = x_1 \wedge \cdots \wedge x_n$$

sends $x_1^* \wedge \cdots \wedge x_k^*$ to a decomposable $(n-k)$-vector of V. The same holds for h_w, since it is a scalar multiple of $h_{w'}$. $\qquad\square$

It follows from Proposition 6.1 that every $(n-1)$-vector of V is decomposable. In the case when $1 < k < n-1$, there are non-decomposable k-vectors. We refer [Sternberg (1983), Section 1.5] for examples and a criterion of decomposability.

Let $x_1, \ldots, x_n$ be a base of V. Suppose that $x_1^*, \ldots, x_n^*$ are the vectors of the dual base of V^* and $x_i^* \cdot x_j = \delta_{ij}$. The associated n-vector is $w = x_1 \wedge \cdots \wedge x_n$ and we write e_w for the linear isomorphism of V to V^* transferring every x_i to x_i^*. The composition of $\wedge^k(e_w)$ and h_w is a linear isomorphism of $\wedge^k(V)$ to $\wedge^{n-k}(V)$ for every $k \in \{1, \ldots, n-1\}$. It is called the *Hodge star isomorphism* corresponding to the n-vector w and denoted by $*_w$. In the case when $n = 2k$, we get a linear automorphism of $\wedge^k(V)$ whose square coincides with $(-1)^k \mathrm{Id}_V$ (see the proof of Proposition 6.1), i.e this is an involution only for even k.

Now we show that there exist linear isomorphisms $s : V \to V^*$ such that $h_w \wedge^k(s)$, $w \in \wedge^n(V)$ is an involution if $n = 2k$ and k is odd.

Example 6.2. Let $x_1, \ldots, x_n$ and $y_1, \ldots, y_n$ be bases of V. Consider the associated n-vectors

$$w = x_1 \wedge \cdots \wedge x_n \quad \text{and} \quad v = y_1 \wedge \cdots \wedge y_n.$$

Suppose that $x_1^*, \ldots, x_n^*$ and $y_1^*, \ldots, y_n^*$ are the vectors of the corresponding dual bases of V^* such that $x_i^* \cdot x_j = \delta_{ij}$ and $y_i^* \cdot y_j = \delta_{ij}$. Let s be the linear isomorphism of V to V^* transferring every x_i to y_i^*. An easy verification shows that $s^* : V \to V^*$ sends each y_i to x_i^*. Thus the contragradient

$\check{s} : V^* \to V$ is the linear isomorphism transferring every x_i^* to y_i. This implies that

$$h_v \wedge^k (s) = \wedge^{n-k}(\check{s}e_w)*_w$$

for every $k \in \{1, \ldots, n-1\}$. We rewrite this equality as follows

$$h_w \wedge^k (s) = \det(M) \wedge^{n-k} (\check{s}e_w)*_w,$$

where M is the matrix of decomposition of $x_1, \ldots, x_n$ in the base $y_1, \ldots, y_n$. If $n = 2k$ then the square of $h_w \wedge^k (s)$ is equal to

$$\det(M)h_w \wedge^k (s\check{s}e_w) *_w .$$

The latter linear automorphism is identity if $s = s^*$ and $\det(M) = (-1)^k$. Semilinear isomorphisms $s : V \to V^*$ satisfying these conditions exist.

6.2   Grassmannians

As in the previous section, we suppose that V is an n-dimensional vector space over a field F.

Lemma 6.4. *Linearly independent vectors* $x_1, \ldots, x_k$ *and* $y_1, \ldots, y_k$ *span the same k-dimensional subspace of V if and only if the associated k-vectors*

$$x_1 \wedge \cdots \wedge x_k \quad and \quad y_1 \wedge \cdots \wedge y_k$$

are linearly dependent.

Proof. If $x_1, \ldots, x_k$ and $y_1, \ldots, y_k$ span the same k-dimensional subspace then

$$y_1 \wedge \cdots \wedge y_k = \det(M) \, x_1 \wedge \cdots \wedge x_k,$$

where M is the matrix of decomposition of $y_1, \ldots, y_k$ in the base $x_1, \ldots, x_k$.

Suppose that $x_1, \ldots, x_k$ and $y_1, \ldots, y_k$ span distinct k-dimensional subspaces S and U. There is a base of V such that both S and U are spanned by subsets of this base. Then the associated base of $\wedge^k(V)$ contains scalar multiples of our k-vectors which means these k-vectors are linearly independent. $\square$

Following [Hirschfeld and Thas (1991); Hodge and Pedoe (1952)] we consider the *Plücker mapping*

$$g : \mathcal{G}_k(V) \to \mathcal{G}_1(\wedge^k(V))$$

which sends every k-dimensional subspace $S \subset V$ to the 1-dimensional subspace of $\wedge^k(V)$ containing $x_1 \wedge \cdots \wedge x_k$, where $x_1, \ldots, x_k$ form a base of S. By Lemma 6.4, this mapping is well-defined and injective. Since the image of the Plücker mapping consists of all 1-dimensional subspaces of $\wedge^k(V)$ spanned by decomposable k-vectors, the Plücker mapping is not surjective if $1 < k < n - 1$.

Lemma 6.5. *The Plücker mapping transfers every line of $\mathcal{G}_k(V)$ to a line of $\Pi_{\wedge^k(V)}$. If a line of $\Pi_{\wedge^k(V)}$ is contained in the image of the Plücker mapping then this line is the image of a certain line of $\mathcal{G}_k(V)$.*

Proof. Easy verification. $\square$

In what follows we will suppose that $1 < k < n - 1$ and identify the image of the Plücker mapping with the Grassmannian $\mathcal{G}_k(V)$.

All linear isomorphisms

$$h_{v^*} : \wedge^k(V) \to \wedge^{n-k}(V^*), \quad v^* \in \wedge^n(V^*)$$

define the same collineation of $\Pi_{\wedge^k(V)}$ to $\Pi_{\wedge^{n-k}(V^*)}$ (since these linear isomorphisms are coincident up to a scalar multiple). By Proposition 6.1, this collineation transfers $\mathcal{G}_k(V)$ to $\mathcal{G}_{n-k}(V^*)$; moreover, the restriction of this collineation to $\mathcal{G}_k(V)$ coincides with the annihilator mapping (see the proof of Proposition 6.1).

The apartment of $\mathcal{G}_k(V)$ defined by a base $x_1, \ldots, x_n$ of V can be considered as the base of the projective space $\Pi_{\wedge^k(V)}$ whose elements contain the k-vectors of type

$$x_{i_1} \wedge \cdots \wedge x_{i_k}, \quad i_1 < \cdots < i_k.$$

There is a similar interpretation for $J(n, k)$-subsets of $\mathcal{G}_k(V)$. Note that the $J(n, k)$-subset $J_k(\mathcal{X})$, where $\mathcal{X}$ is an m-independent subset of Π_V consisting of n elements and $m \geq 2k$, need not to be an m-independent subset of $\Pi_{\wedge^k(V)}$, see Example 6.1.

Example 6.3. If u is a semilinear automorphism of V then $\wedge^k(u)$ is a semilinear automorphism of $\wedge^k(V)$ and $(\wedge^k(u))_1$ is a collineation of $\Pi_{\wedge^k(V)}$ to itself. The restriction of this collineation to $\mathcal{G}_k(V)$ coincides with $(u)_k$.

Example 6.4. If $s : V \to V^*$ is a semilinear isomorphism then the composition of $\wedge^k(s)$ and

$$h_v : \wedge^k(V^*) \to \wedge^{n-k}(V), \quad v \in \wedge^n(V)$$

is a semilinear isomorphism of $\wedge^k(V)$ to $\wedge^{n-k}(V)$ and we get a semilinear automorphism of $\wedge^k(V)$ in the case when $n = 2k$. The associated
collineation of $\Pi_{\wedge^k(V)}$ to $\Pi_{\wedge^{n-k}(V)}$ does not depend on $v \in \wedge^n(V)$ and its
restriction to $\mathcal{G}_k(V)$ coincides with $(s)_k^*$.

There is the following interpretation of Chow's theorem in terms of
semilinear automorphisms of exterior powers.

Theorem 6.1. *If l is a semilinear automorphism of $\wedge^k(V)$ such that $(l)_1$
transfers $\mathcal{G}_k(V)$ to itself then one of the following possibilities is realized:*

- *l is a scalar multiple of $\wedge^k(u)$, where u is a semilinear automorphism
of V;*
- *$n = 2k$ and l is the composition of $\wedge^k(s)$ and h_v, where $v \in \wedge^n(V)$ and
$s : V \to V^*$ is a similinear isomorphism.*

Proof. Stars and tops of $\mathcal{G}_k(V)$ are subspaces of the projective space
$\Pi_{\wedge^k(V)}$. By the second part of Lemma 6.5, every maximal subspace of
$\Pi_{\wedge^k(V)}$ contained in $\mathcal{G}_k(V)$ is a star or a top. This implies that $(l)_1$ preserves the class of maximal cliques of $\Gamma_k(V)$. Hence the restriction of $(l)_1$
to $\mathcal{G}_k(V)$ is an automorphism of $\Gamma_k(V)$. By Chow's theorem, for every
automorphism of $\Gamma_k(V)$ one of the following possibilities is realized: it is
induced by an automorphism of V or $n = 2k$ and the automorphism is
induced by a semilinear isomorphism of V to V^*.

Consider the case when the restriction of $(l)_1$ to $\mathcal{G}_k(V)$ is the automorphism of $\Gamma_k(V)$ induced by a semilinear automorphism u of V. Then
the restrictions of $(l)_1$ and $(\wedge^k(u))_1$ to $\mathcal{G}_k(V)$ are coincident. Let $\mathcal{X}$ be a
maximal clique of $\Gamma_k(V)$. Denote by $U(\mathcal{X})$ the corresponding subspace of
the vector space $\wedge^k(V)$. In the case when $\mathcal{X}$ is a star $[S)_k$, the subspace
$U(\mathcal{X})$ consists of all k-vectors $x \wedge x_1 \wedge \cdots \wedge x_{k-1}$, where $x_1, \ldots, x_{k-1}$ form
a base of S. If $\mathcal{X}$ is a top $\langle U]_k$ then $U(\mathcal{X})$ coincides with $\wedge^k(U)$. Since
the restrictions of $(l)_1$ and $(\wedge^k(u))_1$ to $\mathcal{X}$ are coincident, Proposition 1.6
implies the existence of a non-zero scalar $a_{\mathcal{X}}$ such that

$$l|_{U(\mathcal{X})} = a_{\mathcal{X}} \wedge^k (u)|_{U(\mathcal{X})}.$$

If $\mathcal{X}$ and $\mathcal{Y}$ are maximal cliques of $\Gamma_k(V)$ and the corresponding vertices of
the graph $\mathrm{Cl}_k(V)$ are adjacent then

$$\dim(U(\mathcal{X}) \cap U(\mathcal{Y})) = 2$$

which implies that $a_{\mathcal{X}} = a_{\mathcal{Y}}$. The connectedness of the graph $\mathrm{Cl}_k(V)$
guarantees that the equality $a_{\mathcal{X}} = a_{\mathcal{Y}}$ holds for any pair of maximal cliques

$\mathcal{X}$ and $\mathcal{Y}$. The union of all subspaces $U(\mathcal{X})$ coincides with the set D formed by all decomposable k-vectors. Thus

$$l|_D = a \wedge^k (u)|_D$$

for a certain non-zero scalar a. Since D contains a base of $\wedge^k(V)$ (we can take any base of $\wedge^k(V)$ obtained from a base of V), we have $l = a \wedge^k (u)$.

Suppose that $n = 2k$ and the restriction of $(l)_1$ to $\mathcal{G}_k(V)$ is the automorphism of $\Gamma_k(V)$ induced by a semilinear isomorphism $s : V \to V^*$. Then the restrictions of $(l)_1$ and $(h_v \wedge^k (s))_1$, $v \in \wedge^n(V)$ to $\mathcal{G}_k(V)$ are coincident. Using the same arguments, we show that l is a scalar multiple of $h_v \wedge^k (s)$. This implies that l coincides with $h_{v'} \wedge^k (s)$ for a certain $v' \in \wedge^n(V)$. $\qquad\square$

If the field F is algebraically closed then for every $a \in F$ there exists $b \in F$ such that $b^k = a$ and for every semilinear mapping $l : V \to V$ we have

$$a \wedge^k (l) = \wedge^k (bl).$$

This observation together with Theorem 6.1 give the following.

Corollary 6.1 ([Westwick (1964)]). *Suppose that l is a semilinear automorphism of $\wedge^k(V)$ such that $(l)_1$ transfers $\mathcal{G}_k(V)$ to itself. If the field F is algebraically closed then one of the following possibilities is realized:*

- *$l = \wedge^k(u)$, where u is a semilinear automorphism of V;*
- *$n = 2k$ and l is the composition of $\wedge^k(s)$ and h_v, where $v \in \wedge^n(V)$ and $s : V \to V^*$ is a similinear isomorphism.*

Now we describe the group of all semilinear automorphisms of $\wedge^k(V)$ preserving $\mathcal{G}_k(V)$. Denoted by $\Gamma\mathrm{L}(V)_k$ the quotient group of $\Gamma\mathrm{L}(V)$ by the subgroup of all homothetic transformations $x \to ax$, where $a \in F$ satisfies $a^k = 1$. A semilinear automorphism of V induces the identity transformation of $\wedge^k(V)$ if and only if it is such a homothety.

First we consider the case when F is algebraically closed.

Corollary 6.2. *Suppose that F is algebraically closed. If $n \neq 2k$ then the group of all semilinear automorphisms of $\wedge^k(V)$ preserving $\mathcal{G}_k(V)$ is isomorphic to $\Gamma\mathrm{L}(V)_k$. In the case when $n = 2k$, this group can be presented as the semidirect product $\Gamma\mathrm{L}(V)_k \rtimes \mathbb{Z}_2$.*

Proof. The first statement is obvious and we consider the case $n = 2k$. Let G be the group of all semilinear automorphisms of $\wedge^k(V)$ induced by semilinear automorphisms of V. This group is isomorphic to $\Gamma\mathrm{L}(V)_k$. We take any semilinear isomorphism $s : V \to V^*$ and any n-vector $w \in \wedge^n(V)$. Denote by l the composition of $\wedge^k(s)$ and h_w. Then every element of lGl^{-1} belongs to G, since it induces an automorphism of $\Gamma_k(V)$ transferring stars to stars and tops to tops. This means that G is a normal subgroup. By Example 6.2, there exists a linear isomorphism $s : V \to V^*$ such that $h_w \wedge^k (s)$, $w \in \wedge^n(V)$ is an involution. Our group is the semidirect product of G and the group of order 2 generated by $h_w \wedge^k (s)$. $\qquad \square$

Let $F^\times$ be the multiplicative group of the field F. Consider the homomorphism of the direct product $F^\times \times \Gamma L(V)$ to $\Gamma L(\wedge^k(V))$ transferring every pair (a, l) to $a\wedge^k(l)$. The kernel of this homomorphism is formed by all pairs (a^k, u_a), where u_a is the homothety transferring every $x \in V$ to $a^{-1}x$. In the case when $n = 2k$, there is a similar homomorphism of $F^\times \times G$ to $\Gamma L(\wedge^k(V))$, where G is the group of all semilinear automorphisms of $\wedge^k(V)$ obtained from semilinear automorphisms of V and semilinear isomorphisms of V to V^*.

Corollary 6.3. *The group of all semilinear automorphisms of $\wedge^k(V)$ preserving $\mathcal{G}_k(V)$ is isomorphic to*

$$(F^\times \times \Gamma L(V))/\{(a^k, u_a) : a \in F^\times, u_a(x) = a^{-1}x\}$$

if $n \neq 2k$. In the case when $n = 2k$, this group is isomorphic to

$$(F^\times \times (\Gamma L(V)_k \rtimes \mathbb{Z}_2))/\{(a^k, \wedge^k(u_a)) : a \in F^\times, u_a(x) = a^{-1}x\}.$$

Suppose that V' is an n'-dimensional vector space over a field F'. Consider the following generalizations of Examples 6.3 and 6.4.

Example 6.5. Every semilinear mapping $l : V \to V'$ induces the semilinear mapping

$$\wedge^k(l) : \wedge^k(V) \to \wedge^k(V').$$

If l is a semilinear m-embedding and $m \geq k$ then the restriction of $(\wedge^k(l))_1$ to $\mathcal{G}_k(V)$ coincides with $(l)_k$, but $\wedge^k(l)$ is not necessarily a semilinear m-embedding (Example 6.1).

Example 6.6. For any semilinear mapping $s : V \to V'^*$ the composition of $\wedge^k(s)$ and

$$h_v : \wedge^k(V'^*) \to \wedge^{n'-k}(V'), \quad v \in \wedge^{n'}(V')$$

is a semilinear mapping of $\wedge^k(V)$ to $\wedge^{n'-k}(V')$. In the case when s is a semilinear m-embedding and $m \geq k$, the restriction of $(h_v \wedge^k (s))_1$ to $\mathcal{G}_k(V)$ coincides with $(s)_k^*$. Note that the mapping $(h_v \wedge^k (s))_1$ does not depend on $v \in \wedge^{n'}(V')$.

We define the distance between two decomposable k-vectors v and w as the smallest number i such that there exist vectors $x_1, \ldots, x_{k-i} \in V$ satisfying
$$v = x_1 \wedge \cdots \wedge x_{k-i} \wedge y_1 \wedge \cdots \wedge y_i \quad \text{and} \quad w = x_1 \wedge \cdots \wedge x_{k-i} \wedge z_1 \wedge \cdots \wedge z_i.$$
This distance is equal to the distance between the k-dimensional subspaces of V corresponding to v and w. Theorem 3.1 gives the following.

Corollary 6.4. *Suppose that l is a semilinear mapping of $\wedge^k(V)$ to $\wedge^k(V')$ which transfers decomposable k-vectors to decomposable k-vectors and preserves the distance between them. If $k \leq n - k$ the one of the following possibilities is realized:*

- *l is a scalar multiple of $\wedge^k(u)$, where $u : V \to V'$ is a semilinear m-embedding and $m \geq 2k$;*
- *there exist $U \in \mathcal{G}_{2k}(V')$ and a semilinear $(2k)$-embedding $s : V \to U^*$ such that l is the composition of $\wedge^k(s)$ and h_v, $v \in \wedge^{2k}(U)$.*

Proof. The mapping $(l)_1$ transfers $\mathcal{G}_k(V)$ to a subset of $\mathcal{G}_k(V')$ and its restriction to $\mathcal{G}_k(V)$ is an isometric embedding of $\Gamma_k(V)$ in $\Gamma_k(V')$. Since $k \leq n - k$, Theorem 3.1 implies that this restriction is induced by a semilinear m-embedding $u : V \to V'$ with $m \geq 2k$ or it can be obtained from a semilinear $(2k)$-embedding $s : V \to U^*$, where $U \in \mathcal{G}_{2k}(V')$. So,
$$(l)_1|_{\mathcal{G}_k(V)} = (\wedge^k(u))_1|_{\mathcal{G}_k(V)}$$
or
$$(l)_1|_{\mathcal{G}_k(V)} = (h_v \wedge^k (s))_1|_{\mathcal{G}_k(V)}, \quad v \in \wedge^{2k}(U).$$
As in the proof of Theorem 6.1, we establish that l is a scalar multiple of $\wedge^k(u)$ or $h_v \wedge^k (s)$. In the second case, l coincides with $h_{v'} \wedge^k (s)$. $\square$

Remark 6.1. If $k > n - k$ then we can apply the above arguments to the composition of the Hodge star isomorphism of $\wedge^{n-k}(V)$ to $\wedge^k(V)$ and l (we leave all details to the readers).

Using Theorem 5.1 and Corollary 5.1, we establish the following.

Corollary 6.5. *Suppose that l is a semilinear mapping of $\wedge^k(V)$ to $\wedge^k(V')$ such that the mapping $(l)_1$ transfers every apartment of $\mathcal{G}_k(V) \subset \mathcal{G}_1(\wedge^k(V))$ to a $J(n, k)$-subset of $\mathcal{G}_k(V') \subset \mathcal{G}_1(\wedge^k(V'))$. If $k \leq n-k$ then l is one of the mappings described in Corollary 6.4. If $n = n'$ and $(l)_1$ transfers a certain apartment of $\mathcal{G}_k(V)$ to an apartment of $\mathcal{G}_k(V')$ then one of the following possibilities is realized [1]:*

---

[1] We do not require that $k \leq n - k$ in this case.

- *l is a scalar multiple of $\wedge^k(u)$, where $u : V \to V'$ is a strong semilinear embedding;*
- *$n = 2k$ and l is the composition of $\wedge^k(s)$ and h_v, where $v \in \wedge^n(V')$ and $s : V \to V^*$ is a strong semilinear embedding.*

Remark 6.2. Suppose that l is a semilinear mapping of $\wedge^k(V)$ to $\wedge^k(V')$ which transfers decomposable k-vectors to decomposable k-vectors and non-decomposable k-vectors to non-decomposable k-vectors. Then $(l)_1$ sends $\mathcal{G}_k(V)$ to a subset of $\mathcal{G}_k(V')$ and $\mathcal{G}_1(\wedge^k(V)) \setminus \mathcal{G}_k(V)$ to a subset of $\mathcal{G}_1(\wedge^k(V')) \setminus \mathcal{G}_k(V')$. This means that the restriction of $(l)_1$ to $\mathcal{G}_k(V)$ is an embedding of $\Gamma_k(V)$ in $\Gamma_k(V')$. We cannot state that this embedding is isometric.

6.3   Grassmann codes

In this section, we suppose that V is an n-dimensional vector space over the finite field $F = \mathrm{GF}(q)$ consisting of q elements. Then $\mathcal{G}_k(V)$ contains precisely

$$\begin{bmatrix} n \\ k \end{bmatrix}_q = \frac{(q^n - 1)(q^n - q) \dots (q^n - q^{k-1})}{(q^k - 1)(q^n - q) \dots (q^k - q^{k-1})}$$

elements. Suppose that $1 < k < n-1$. Since $\mathcal{G}_k(V)$ is a subset of $\mathcal{G}_1(\wedge^k(V))$, it defines the equivalence class of linear $[l, m]_q$ codes, where

$$l = \begin{bmatrix} n \\ k \end{bmatrix}_q \quad \text{and} \quad m = \binom{n}{k},$$

see Section 2.7. The elements of this class are known as the *Grassmann codes* [Nogin (1996)] (see [Ryan 1 (1987); Ryan 2 (1987)] for the case $k = 2$). Suppose that

$$P_1, \dots, P_l \in \mathcal{G}_1(\wedge^k(V))$$

are the elements of $\mathcal{G}_k(V)$ and denote by U the vector space dual to $\wedge^k(V)$. We take any non-zero k-vectors

$$v_1 \in P_1, \dots, v_l \in P_l$$

and consider the linear injection $u : U \to F^l$ defined as follows

$$u(v^*) = (v^* \cdot v_1, \dots, v^* \cdot v_l)$$

for every $v^* \in U$. Then $C = u(U)$ is one of the Grassmann codes. From this moment we will consider u as a linear isomorphism of U to C. By Lemma 2.22, the contragradient

$$\check{u} : \wedge^k(V) \to C^*$$

transfers $\mathcal{G}_k(V)$ to the projective system $\mathcal{P}(C)$.

Let G be the automorphism group of the code C. By Proposition 1.10, it is isomorphic to the group $\check{G}$ formed by the contragradients of automorphisms of C. Lemma 2.20 shows that a semilinear automorphism s of C^* belongs to $\check{G}$ if and only if it preserves $\mathcal{P}(C)$. The latter is equivalent to the fact that $\check{u}^{-1}s\check{u}$ is a semilinear automorphism of $\wedge^k(V)$ preserving $\mathcal{G}_k(V)$. Thus $\check{G}$ is isomorphic to the group of all semilinear automorphisms of $\wedge^k(V)$ preserving $\mathcal{G}_k(V)$.

So, the automorphism group of every Grassmann code defined by $\mathcal{G}_k(V)$ is isomorphic to the group of all semilinear automorphisms of $\wedge^k(V)$ preserving $\mathcal{G}_k(V)$. The latter group is described in Corollary 6.3 and we refer [Ghorpade and Kaipa (2013)] for more information.

An interesting class of linear codes related to polar Grassmannians can be found in [Cardinali and Giuzzi (2013)].

Bibliography

Artin, E. (1957). *Geometric Algebra* (Interscience Publishers, New York-London).

Baer, R. (1952). *Linear Algebra and Projective Geometry* (Academic Press, New York).

Bachman, G. (1964). *Introduction to p-adic numbers and valuation theory* (Academic Press, New York).

Blunck, A. and Havlicek, H. (2005). On bijections that preserve complementarity of subspaces, *Discrete Math.* **301**, 1, pp. 46–56.

Bogart, K., Goldberg, D. and Gordon, J. (1978). An elementary proof of the MacWilliams theorem on equivalence of codes, *Inform. and Control* **37**, 1, pp. 19–22.

Brauner, H. (1988). Über die von Kollineationen projektiver Räume induzierten Geradenabbildungen, *Sb. österr. Akad. Wiss., Abt. II, math. phys. techn. Wiss.* **197**, pp. 327–332.

Brezuleanu, A. and Rădulescu, D.-C. (1984). About full or injective lineations, *J. Geom.* **23**, 1, pp. 45–60.

Brouwer, A. E., Cohen, A. M. and Neumaier, A. (1989). *Distance-regular graphs*, Ergebnisse der Mathematik und ihrer Grenzgebiete/Results in Mathematics and Related Areas **18** (Springer, Berlin).

Buekenhout, F. (1965). Une généralisation du théorème de von Staudt–Hua, *Acad. Roy. Belg. Bull. Cl. Sci.* **51**, 5, pp. 1282–1293.

Buekenhout, F. and Cohen, A. M. (2013). *Diagram geometry*, Ergebnisse der Mathematik und ihrer Grenzgebiete/Results in Mathematics and Related Areas **57** (Springer, Heidelberg).

Cardinali, I. and Giuzzi, L. (2013). Codes and caps from orthogonal Grassmannians *Finite Fields Appl.* **24**, pp. 148–169.

Ceccherini, P. V. (1967). Collineazioni e semicollineazioni tra spazi affini o proiettivi, *Rend. Mat. Appl.* **26**, pp. 309–348.

Chow, W. L. (1949). On the geometry of algebraic homogeneous spaces, *Ann. of Math.* **50**, 1, pp. 32–67.

Cohn, P. M. (1995). *Skew fields. Theory of general division rings*, Encyclopedia of Mathematics and its applications **57** (Cambridge University Press, Cambridge).

Cojan, S. P. (1985). A generalization of the theorem of von Staudt–Hua–
   Buekenhout, *Mathematica (Cluj)* **27(52)**, 2, pp. 93–96.

Cooperstein, B. N. and Shult, E. E. (1997). Frames and bases of Lie incidence
   geometries, *J. Geom.* **60**, 1–2, pp. 17–46.

Cooperstein, B. N., Kasikova, A. and Shult, E. E. (2005). Witt-type theorems for
   Grassmannians and Lie incidence geometries, *Adv. Geom.* **5**, 1, pp. 15–36.

Cooperstein, B. N. (2013). Witt-Type theorems for subspaces of Lie geometries: A
   survey, in *Groups of Exceptional Type, Coxeter Groups and Related Geometries*, Springer Proceedings in Mathematics and Statistics **82**, pp. 123–133.

van Dam, E. R. and Koolen, J. H. (2005). A new family of distance-regular graphs
   with unbounded diameter, *Inventiones Math.* **162**, 1, pp. 189–193.

Dempwolff, U. (1990). Normal forms and fixed subspaces of semilinear maps, *Boll.
   Un. Math. Ital.* **4-A**, pp. 209-218.

Dempwolff, U. (1999). Normalformen semilinearer Operatoren, *Math. Semester-
   berichte* **46**, pp. 205–214.

Dempwolff, U., Fisher, J. C. and Herman, A. (2000). Semilinear transformations
   over finite fields are Frobenius maps, *Glasgow Math. J.* **42**, 2, pp. 289–295.

Dempwolff, U. (2010). On irreducible semilinear transformations, *Forum Math.*
   22, 6, pp. 1193–1206.

De Schepper, A and Van Maldeghem H. (2014). Graphs, defined by Weyl distance
   or incidence, that determine a vector space, *Linear Algebra Appl.* **449**, pp.
   135 164.

Deza, M. and Laurent, M. (1997). *Geometry of cuts and metrics*, Algorithms and
   Combinatorics **15** (Springer, Berlin).

Dicks, W. and Hartley, B. (1991). On homomorphisms between special linear
   groups over division rings, *Comm. Algebra* **19**, 7, pp. 1919–1943. Errata
   and addenda, *Comm. Algebra* **22**, 15, pp. 6493–6494.

Dieudonné, J. (1971). *La géométrie des groupes classiques* (Springer, New York-
   Heidelberg-Berlin).

Fan, Y., Liu, H. and Puig, L. (2003). Generalized hamming weights and equiva-
   lences of codes, *Sci. China (Series A)*, **46**, 5, pp. 690–695.

Faure, C. A. and Frölicher, A. (1994). Morphisms of projective geometries and
   semilinear maps, *Geom. Dedicata* **53**, 3, pp. 237–262.

Faure, C. A. and Frölicher, A. (2000). *Modern projecive geometry*, Mathematics
   and its Applications **521** (Kluwer Academic Publishers, Dortrecht-Boston-
   London).

Faure, C. A. (2002). An elementary proof of the Fundamental Theorem of Projective Geometry, *Geom. Dedicata* **90**, 1, pp. 145–151.

Fujisaki, T., Koolen, J. H. and Tagami, M. (2006). Some properties of the twisted
   Grassmann graphs, *Innov. Incidence Geom.* **3**, pp. 81–87.

Ghorpade, S. R. and Kaipa, K. V. (2013). Automorphism groups of Grassmann
   codes, *Finite Fields Appl.* **23**, pp. 80–102.

Havlicek, H. (1994). A generalization of Brauner's theorem on linear mappings,
   Mitt. Math. Sem. Univ. Giessen **215**, pp. 27–41.

Havlicek, H. (1995). On Isomorphisms of Grassmann Spaces, *Mitt. Math. Ges
   Hamburg.* **14**, pp. 117–120.

Havlicek, H. and Pankov, M. (2005). Transformations on the product of Grassmann spaces, *Demonstratio Math.* **38**, 3, pp. 675–688.

Hirschfeld, J. W. P. and Thas, J. A. (1991). *General Galois Geometries* (Oxford University Press, Oxford).

Hodge, W. V. D. and Pedoe, D. (1952). *Methods of Algebraic Geometry* (Cambridge University Press, Cambridge).

Hua, L.-K. (1949). On the automorphism of a sfield, *Proc. N.A.S.* **35**, 7, pp. 386–389.

Hua, L.-K. (1951). A theorem on matrices over sfield and its applications, *Acta Math. Sinica* **1**, 2, pp. 109–163.

Huang, W.-l. and Kreuzer, A. (1995). Basis preserving maps of linear spaces, *Arch. Math. (Basel)* **64**, 6, pp. 530–533.

Huang, W.-l. (1998). Adjacency preserving transformations of Grassmann spaces, *Abh. Math. Sem. Univ. Hamburg* 68, pp. 65–77.

Huang, W.-l. and Havlicek, H. (2008). Diameter preserving surjections in the geometry of matrices, *Linear Algebra Appl.* **429**, 1, pp. 376–386.

Huang, W.-l. (2010). Bounded distance preserving surjections in the geometry of matrices, *Linear Algebra Appl.* **433**, 11–12, pp. 1973–1987.

Huang, W.-l. (2011). Bounded distance preserving surjections in the projective geometry of matrices, *Linear Algebra Appl.* **435**, 1, pp. 175–185.

James, D. G. (1982). Projective homomorphisms and von Staudts theorem, *Geom. Dedicata* **13**, 3, pp. 291–294.

Kasikova, A. (2007). Characterization of some subgraphs of point-collinearity graphs of building geometries, *European J. Combin.* **28**, 5, pp. 1493–1529.

Kasikova, A. (2009). Characterization of some subgraphs of point-collinearity graphs of building geometries II, *Adv. Geom.* **9**, 1, pp. 45–84.

Kasikova, A. (2013). A characterization of point shadows of residues in building geometries, *Innov. Incidence Geom.* **13**, pp. 179–206.

Klotzek, B. (1988). Eine Verallgemeinerung des Satzes von v. Staudt–Hua, *Wiss. Z. Pädagog. Hochsch. "Karl Liebknecht" Potsdam* **32**, 1, pp. 169–172.

Kosiorek, J., Matras, A. and Pankov, M. (2008). Distance preserving mappings of Grassmann graphs, *Beiträge Algebra Geom.* **49**, 1, pp. 233–242.

Kreuzer, A. (1998). Projective embeddings of projective spaces, *Bull. Belg. Math. Soc. Simon Stevin* **5**, 2–3, pp. 363–372.

Kreuzer, A. (1998). On isomorphisms of Grassmann spaces, *Aequationes Math.* **56**, 3, pp. 243–250.

Kwiatkowski, M. and Pankov, M. (2015). Isometric embeddings of polar Grassmannians and metric characterizations of their apartments, arXiv:1502.02873, preprint.

Lang, S. (2002). *Algebra*, Graduate Texts in Mathematics **211**, (Springer, New York).

Liu, Z., Wu, X. W., Luo, Y. and Chen, W. (2011). New code equivalence based on relative generalized Hamming weights, *Inform. Sci.* **181**, 19, pp. 4309–4317.

Liu, Z. and Zeng, X. (2013). Further results on the semilinear equivalence of linear codes, *Inform. Sci.* **221**, pp. 571–578.

Lim, M. H. (2010). Surjections on Grassmannians preserving pairs of elements with bounded distance, *Linear Algebra Appl.* **432**, 7, pp. 1703–1707.

MacWilliams, F. J. (1961). *Combinatorial problems of elementary abelian groups*, PhD thesis, Harvard University, Cambridge, MA.

Metsch, K. (1995). A characterization of Grassmann graphs, *European. J. Comb.* **16**, 6, pp. 639–644.

Nogin, D. (1996). Codes associated to Grassmannians, in *Arithmetic, Geometry and Coding Theory* (de Gruyter, Berlin), pp. 145–154.

O'Meara, O. T. (1974). *Lectures on linear groups* (Amer. Math. Soc., Providence).

Pankov, M. (2010). *Grassmannians of classical buildings*, Algebra and Discrete Math. Series **2** (World Scientific, Singapore).

Pankov, M. (2011). Isometric embeddings of Johnson graphs in Grassmann graphs, *J. Algebraic Combin.* **33**, 4, pp. 555–570.

Pankov, M. (2011). Metric characterization of apartments in dual polar spaces, *J. Combin. Theory Ser. A* **118**, 4, pp. 1313–1321.

Pankov, M. (2012). Characterization of apartments in polar Grassmannians, *Bull. Belg. Math. Soc. Simon Stevin* **19**, 2, pp. 345–366.

Pankov, M. (2012). Embeddings of Grassmann graphs, *Linear Algebra Appl.* **436**, 9, pp. 3413–3424.

Pankov, M. (2013). Characterizations of strong semilinear embeddings in terms of general linear and projective linear groups, *Linear Multilinear Algebra* **61**, 11, 1555–1567.

Pankov, M. (2014). Characterization of isometric embeddings of Grassmann graphs, *Adv. Geom.* **14**, 1, pp. 91–108.

Pankov, M. (2014). Isometric embeddings of half-cube graphs in half-spin Grassmannians, *The Electronic Journal of Combinatorics* **21**, 4, P4.4.

Pasini, A. (1994). *Diagram Geometries*, Oxford Science Publications (Clarendon/Oxford University Press, New York).

Porteous, I. R. (1981). *Topological Geometry* (Cambridge University Press, Cambridge).

Ryan, C. T. (1987). An Application of Grassmanian Varieties to Coding Theory, *Congr. Numer.* **57**, pp. 257–271.

Ryan, C. T. (1987). Projective Codes Based on Grassman Varieties, *Congr. Numer.* **57**, pp. 273–279.

Šemrl, P. (2014). The optimal version of Hua's fundamental theorem of geometry of rectangular matrices, *Memoirs Amer. Math. Soc.* **232**, 1089.

Shult, E. E. (2011). *Points and Lines. Characterizing the Classical Geometries*, Universitext (Springer, Berlin).

Sternberg, S. (1983). *Lectures on Differential Geometry* (Chelsea, New York).

Tits, J. (1974). *Buildings of spherical type and finite BN-pairs*, Lecture Notes in Mathematics **386** (Springer, Berlin-New York).

Tsfasman, M., Vlăduţ, S. and Nogin D. (2007). *Algebraic Geometry Codes. Basic notions* (Amer. Math. Soc., Providence).

Ueberberg, J. (2011). *Foundations of incidence geometry. Projective and polar spaces*, Springer Monographs in Mathematics, (Springer, Heidelberg-Berlin).

Ustimenko, V. A., (1978). Maximality of the group $P\Gamma L_n(q)$ that acts on subspaces of dimension m (in Russian). *Dokl. Akad. Nauk SSSR* **240**, 6, pp. 1305–1308.

Wan, Z.-X. (1996). *Geometry of Matrices*, (World Scientific, Singapore).

Ward, H. N. and Wood, J. A. (1996). Characters and the equivalence of codes, *J. Combin. Theory Ser. A* **73**, 2, pp. 348–352.

Westwick, R. (1964). Linear transformations of Grassmann spaces, *Pacific J. Math.* **14**, 3, pp. 1123–1127.

Yale, P. B. (1966). Automorphisms of the complex numbers, *Math. Magazine* **39**, pp. 135–141.

Zha, J. (1996). Determination of homomorphisms between linear groups of the same degree over division rings, *J. Lond. Math. Soc., II. Ser.* **53**, 3, pp. 479–488.

Index

Printed in the United States
By Bookmasters